1 Number recognition

Remember   Accurate use of numbers is important:

Example   Write four thousand and eighty-three in figures = 4083

Exercise 1   Write these numbers in figures:

1a. five   b. eleven   c. seventeen

2a. thirty-six   b. one hundred   c. one hundred and eight

3a. two hundred and fourteen   b. four hundred and sixty-seven

4a. eight hundred and eighty   b. one thousand

5a. one thousand and three   b. one thousand and ninety-two

6a. two thousand one hundred   b. two thousand three hundred and four

7a. three thousand six hundred and sixteen   b. five thousand

8a. five thousand eight hundred and three   b. seven thousand nine hundred and thirty

9a. nine thousand nine hundred and nine   b. ten thousand

10a. ten thousand and ten   b. ten thousand one hundred

Example   Write 3602 in words = **three thousand six hundred and two**

Exercise 2   Write down these numbers in words:

1a. 6   b. 14   c. 20   d. 34

2a. 73   b. 103   c. 130   d. 195

3a. 416   b. 1000   c. 1235   d. 1308

4a. 2604   b. 3012   c. 3599   d. 5400

5a. 6220   b. 6058   c. 8999   d. 10 000

Everyday Maths Practice

R. Christon and P. Newton

Oxford University Press 1982

contents

2 Place value

Remember

The *place* occupied by a figure affects its value:

Examples

1. 1<u>2</u>6   **The 2 has value twenty**

2. <u>3</u>026   **The 3 has value three thousand**

3. 240<u>9</u>   **The 9 has value nine**

Exercise 1

Write down the value of each underlined number in words:

1a. <u>2</u>4	b. <u>2</u>04	c. 4<u>2</u>	d. 4<u>2</u>0
2a. <u>1</u>80	b. 80<u>1</u>	c. <u>1</u>080	d. 8<u>1</u>0
3a. <u>6</u>7	b. <u>6</u>70	c. <u>6</u>07	d. 7<u>6</u>2
4a. 8941	b. 9<u>8</u>41	c. 80<u>9</u>0	d. 841<u>9</u>
5a. 1<u>5</u>	b. 1<u>5</u>151	c. 1<u>8</u>	d. 1<u>8</u>162

In a code, the numbers: 0, 1, 2, 3, 4, 5, 6, 7, 8, 9

are given by:

Write down the value of each underlined symbol in words:

Examples

1. The ☥ has value ninety ✓

2. The ☥ has value four hundred ✓

Exercise 2

1a. fifty seven

1b. forty seven

1c. one hundred and eighty one

1d. 8 hundred and eighty eight

2a. six hundred and seventy

2b. six hundred

2c. eight

2d. thousand

3a. one

3b.

3c.

3d.

4a.

4b.

4c.

4d.

5a.

5b.

5c.

5d.

6a.

6b.

6c.

6d.

5

3 Addition

Examples

63	47	146	418	24326
+25	+38	+ 86	632	17848
88	85	232	+356	+30416
			1406	72590

Exercise 1

1a. 4   b. 24   c. 132   d. 1240
    +5   +31   + 36   +3529

2a. 1   b. 24   c. 142   d. 3284
     4   32   306   1305
    +3   +41   +430   + 410

3a. 6   b. 46   c. 136   d. 1638
    +8   +35   + 48   +2419

4a. 6   b. 26   c. 143   d. 1234
     5   31   234   2308
    +7   +28   +106   +1056

5a. 68   b. 186   c. 3466   d. 1688
    +59   + 58   +1337   +2756

6a. 34   b. 364   c. 3125   d. 1243
     65   182   1684   2305
    +26   + 65   +2065   + 544

7a. 263   b. 3146   c. 1564   d. 4645
     389   2816   2469   3276
    +652   +2683   + 200   +2303

8a. 26 342 b. 31 656   c. 26 136   d. 24 855
     17 202   24 312   16 360   17 639
    +21 214   +16 320   +31 862   +24 388

9. Crowds at three of Liverpool's home games were   49 875
                                                     42 799
                                             and   41 650

                               Find the total attendance   _____

10. In each of three days a long-distance runner ran   6400 metres
                                                         8704 metres
                                              and   5822 metres

                               How far did he run altogether?   _____

11. Add together eight thousand four hundred and thirty two; three thousand and seven; and two thousand four hundred and eight.

12. Record sales for the 'Addition' pop group's single in three weeks were 13 062, 8973 and 15 604. How many records did they sell in this time?

13. At a general election the Conservative candidate received 36 739 votes, the Labour candidate polled 28 620 and the Liberal man received 12 063. How many votes were counted altogether?

14. On a certain Saturday during the football season 'Bigwoods' Pools paid out £75 850 first prize, £40 740 second prize and £18 954 third prize. What was the total amount paid to the three winners?

15. The 'Mathemagics' pop group gave an open air concert and 6875 people came to listen. They were so popular that they gave two more concerts. 35 674 came to the second and 65 974 to the third. What was the grand total of the audiences?

4 Subtraction

Examples

337	132	3463	3046	24 685
− 24	− 18	− 285	− 158	−12 791
313	114	3178	2888	11 894

Exercise 1

1a. 8 − 3   b. 16 − 5   c. 26 − 14   d. 48 − 13

2a. 168 − 35   b. 236 − 114   c. 3642 − 1320   d. 6528 − 3405

3a. 16 − 9   b. 23 − 17   c. 33 − 29   d. 65 − 28

4a. 135 − 17   b. 264 − 48   c. 385 − 166   d. 416 − 208

5a. 1256 − 83   b. 6446 − 173   c. 2288 − 1094   d. 4336 − 2152

6a. 136 − 89   b. 314 − 65   c. 444 − 189   d. 362 − 193

7a. 4354 − 1428   b. 7362 − 3737   c. 7394 − 4749   d. 8393 − 3564

8a. 3657	b. 3641	c. 6867	d. 5646
−1729	−1666	−3978	−2879

9a. 6570	b. 7305	c. 8046	d. 8000
−3796	−3758	−4769	−7862

10a. 4000	b. 6000	c. 8000	d. 10000
−1638	−3869	−6836	− 6271

11a. 36 514	b. 28 365	c. 69 328	d. 41 632
−18 401	−18 442	−39 936	−16 763

12. The 'Bongo' salesman delivers some dog food to a store 88 miles from his home. On his way back, he stops after 32 miles to buy a paper. How far has he to go to reach home?

13. The English Schoolboys Basketball team scored 74 points in a game. How many did they score in the first half if they scored 37 in the second?

14. Subtract three hundred and ninety-six from eight hundred and seventy-two.

15. A cricket team scored 874 runs in a match. If they scored 385 runs in the first innings, how many did they score in the second?

16. In an election the Labour candidate polled 24 629 votes and the Conservative candidate polled 16 330 votes. Who won, and by how many?

17. 53 186 football supporters were watching Manchester United play Arsenal. After Manchester scored their fifth goal, 16 872 Arsenal supporters walked out of the ground. How many were left?

5 Multiplication by 10, 100 and 1000

Examples

1. $3 \times 10 = 30$
2. $100 \times 3 = 300$
3. $10 \times 32 = 320$
4. $14 \times 1000 = 14\,000$

Exercise 1

Work out the following:

1a. 2×10	b. 100×4	c. 2×100	d. 10×4
2a. 16×10	b. 12×100	c. 1000×16	d. 12×10
3a. 63×10	b. 54×100	c. 1000×82	d. 941×10

4a. 10×12 **b.** 100×18 **c.** 100×81 **d.** 4×1000

5a. 10×10 **b.** 1000×10 **c.** 10×100 **d.** 100×100

6. A game of pinball takes 5 minutes to play, on average. How long would it take to play 10 games?

7. A game of pinball costs 10p.
   a. How much would 4 games cost?
   b. How much would 10 games cost?

8. "Ripusoff" fruit machines pay 10 times the cost of a game for the winning line of three bells:
   a. If a game costs 20p, how much would you win if three bells came up?
   b. If a game costs 15p, how much would you win if the three bells came up?

9. It costs £25 per year to service a "Ripusoff" fruit machine.
   a. How much does it cost to service one machine for ten years?
   b. How much do ten machines cost in service bills in ten years?

10. A man takes some money into a bank. He changes £20 for 10p coins so that he can play 'Space Invaders' all weekend. If Space Invaders costs 10p a game, how many games is he able to have?

6 Division by 10, 100 and 1000

Examples

1. $400 \div 10 = 40$ 2. $400 \div 100 = 4$

3. $3800 \div 10 = 380$ 4. $12\,000 \div 1000 = 12$

5. $20p \div 10 = 2p$ 6. $20p \div 2 = 10p$

Exercise 1   Work out the following:

1a. $20 \div 10$ **b.** $40 \div 10$ **c.** $2000 \div 10$ **d.** $2000 \div 100$

2a. $160 \div 10$ **b.** $14\,000 \div 1000$ **c.** $120 \div 10$ **d.** $14\,000 \div 100$

3a. $6000 \div 10$ **b.** $6000 \div 1000$ **c.** $6000 \div 100$ **d.** $8100 \div 100$

4a. $1000 \div 10$ **b.** $1000 \div 100$ **c.** $1000 \div 1000$ **d.** $100 \div 100$

5. A game of pinball costs 10p.
   a. If you have 60p to play pinball, how many games could you have? *6*
   b. If somebody then gives you £1·20, how many more games can you have? *12* ✓

6. a. You win the jackpot on the fruit-machine and collect 800p. How many £1 notes will the bank give you for your winnings? *8* ✓
   b. If, instead, you shared your winnings between ten friends, how many pence would each of them get? *80p* ✓

9

7. Bert travels to and from work each day on the bus.
   a. Bert works a 5 day week. How many journeys does he make on the bus each week?
   b. Bert travels 120 miles a week. How far is each journey on the bus?

8. Angabout the tortoise only moves 10 metres for every 800 metres that you walk.
   a. How far have you walked if Angabout has moved 1 metre?
   b. How far has Angabout travelled if you have walked 8000 metres?

7 Multiplication tables

Do you know your multiplication tables? First, copy the crossword into your book. Then see how long it takes you to complete it:

ACROSS

1. 4 X 6	12. 7 X 4
2. 4 X 9	15. 3 X 5
3. 9 X 9	17. 7 X 3
5. 6 X 7	19. 10 X 5
7. 3 X 6	22. 9 X 10
8. 11 X 12	23. 7 X 7
10. 11 X 6	24. 4 X 5
11. 9 X 8	

DOWN

1. 3 X 9	10. 6 X 10
2. 4 X 8	13. 10 X 8
4. 3 X 4	14. 12 X 1
5. 6 X 8	16. 5 X 11
6. 9 X 7	18. 12 X 12
7. 4 X 4	20. 9 X 11
8. 11 X 11	21. 3 X 10
9. 11 X 2	

8 Multiplication by numbers less than 10

Examples

1.
$$\begin{array}{r} 17 \\ \times\ 6 \\ \hline 102 \\ \hline \end{array}$$

2.
$$\begin{array}{r} 112 \\ \times\ \ 5 \\ \hline 560 \\ \hline \end{array}$$

3.
$$\begin{array}{r} 238 \\ \times\ \ 7 \\ \hline 1666 \\ \hline \end{array}$$

Exercise 1

1a.
$$\begin{array}{r} 9 \\ \times 6 \\ \hline \end{array}$$
b.
$$\begin{array}{r} 8 \\ \times 7 \\ \hline \end{array}$$
c.
$$\begin{array}{r} 12 \\ \times\ 8 \\ \hline \end{array}$$
d.
$$\begin{array}{r} 10 \\ \times\ 9 \\ \hline \end{array}$$

2a.
$$\begin{array}{r} 13 \\ \times\ 3 \\ \hline \end{array}$$
b.
$$\begin{array}{r} 22 \\ \times\ 4 \\ \hline \end{array}$$
c.
$$\begin{array}{r} 31 \\ \times\ 2 \\ \hline \end{array}$$
d.
$$\begin{array}{r} 33 \\ \times\ 3 \\ \hline \end{array}$$

3a.
$$\begin{array}{r} 16 \\ \times\ 3 \\ \hline \end{array}$$
b.
$$\begin{array}{r} 24 \\ \times\ 5 \\ \hline \end{array}$$
c.
$$\begin{array}{r} 36 \\ \times\ 7 \\ \hline \end{array}$$
d.
$$\begin{array}{r} 45 \\ \times\ 9 \\ \hline \end{array}$$

4a.
$$\begin{array}{r} 122 \\ \times\ \ 4 \\ \hline \end{array}$$
b.
$$\begin{array}{r} 231 \\ \times\ \ 3 \\ \hline \end{array}$$
c.
$$\begin{array}{r} 312 \\ \times\ \ 3 \\ \hline \end{array}$$
d.
$$\begin{array}{r} 113 \\ \times\ \ 2 \\ \hline \end{array}$$

5a.
$$\begin{array}{r} 126 \\ \times\ \ 3 \\ \hline \end{array}$$
b.
$$\begin{array}{r} 102 \\ \times\ \ 6 \\ \hline \end{array}$$
c.
$$\begin{array}{r} 114 \\ \times\ \ 7 \\ \hline \end{array}$$
d.
$$\begin{array}{r} 129 \\ \times\ \ 3 \\ \hline \end{array}$$

6. In his will, a farmer left:
$$\begin{array}{r} £270 \\ \times\ \ 8 \\ \hline \end{array}$$ children.
to each of his:

   How much money altogether? _____

Find the money totals in questions 7 to 11:

7. After winning the Pools a Doctor gave: £330
   to each of his: __6__ nurses

8. After finding the treasure, the captain left: 575 doubloons
   to each of his: __8__ crew

9. In his will a Director left: £845
   to each of his: __9__ secretaries

10. A Taxi Firm owner paid: £3400
    for each of his: __5__ new cars

11. For his 66th birthday a millionaire gave: 6666 £1 premium bonds
    to each of his: __4__ nieces

12. What is the cost of six bungalows if each one costs £22 350?

13. Wonderboots United, the local soccer club, promised each of its team a bonus of £650 if they won the local derby with Foulfeet City. They did win, but two of Wonderboots' team were sent off so they didn't get paid. How much did Wonderboots pay in bonuses?

9 Division by numbers less than 10

Remember   When dividing, some questions give an exact answer. Others leave a remainder:

Examples

1. $69 \div 3$

$$\begin{array}{r} 23 \\ 3\,\overline{)69} \end{array}$$

2. $486 \div 2$

$$\begin{array}{r} 243 \\ 2\,\overline{)486} \end{array}$$

3. $144 \div 6$

$$\begin{array}{r} 24 \\ 6\,\overline{)144} \end{array}$$

4. $243 \div 2$

$$\begin{array}{r} 121\ r\ 1 \\ 2\,\overline{)243} \end{array}$$

5. $67 \div 5$

$$\begin{array}{r} 13\ r\ 2 \\ 5\,\overline{)67} \end{array}$$

Exercise 1

1a. $26 \div 2$	b. $54 \div 2$	c. $246 \div 2$	d. $186 \div 2$	e. $312 \div 2$
2a. $36 \div 3$	b. $42 \div 3$	c. $693 \div 3$	d. $249 \div 3$	e. $432 \div 3$
3a. $84 \div 4$	b. $56 \div 4$	c. $488 \div 4$	d. $284 \div 4$	e. $548 \div 4$
4a. $55 \div 5$	b. $75 \div 5$	c. $505 \div 5$	d. $355 \div 5$	e. $265 \div 5$
5a. $60 \div 6$	b. $84 \div 6$	c. $660 \div 6$	d. $366 \div 6$	e. $432 \div 6$
6a. $77 \div 7$	b. $91 \div 7$	c. $700 \div 7$	d. $497 \div 7$	e. $525 \div 7$
7a. $80 \div 8$	b. $96 \div 8$	c. $808 \div 8$	d. $648 \div 8$	e. $688 \div 8$
8a. $99 \div 9$	b. $63 \div 9$	c. $999 \div 9$	d. $729 \div 9$	e. $765 \div 9$
9a. $27 \div 2$	b. $79 \div 2$	c. $285 \div 2$	d. $437 \div 2$	e. $513 \div 2$
10a. $65 \div 3$	b. $47 \div 3$	c. $634 \div 3$	d. $353 \div 3$	e. $436 \div 3$
11a. $47 \div 4$	b. $57 \div 4$	c. $849 \div 4$	d. $457 \div 4$	e. $539 \div 4$
12a. $59 \div 5$	b. $76 \div 5$	c. $558 \div 5$	d. $568 \div 5$	e. $684 \div 5$
13a. $68 \div 6$	b. $74 \div 6$	c. $669 \div 6$	d. $685 \div 6$	e. $739 \div 6$
14a. $79 \div 7$	b. $85 \div 7$	c. $778 \div 7$	d. $786 \div 7$	e. $836 \div 7$
15a. $89 \div 8$	b. $97 \div 8$	c. $889 \div 8$	d. $895 \div 8$	e. $935 \div 8$
16a. $95 \div 9$	b. $275 \div 9$	c. $903 \div 9$	d. $939 \div 9$	e. $526 \div 9$

17. A hard-up Maths teacher leaves £98 in his will to be shared equally amongst his seven children. How much does each child receive?

18. A lollipop factory produces lollipops at a steady rate and has an output of 2968 lollipops in a seven day week. How many lollipops are produced each day?

19. The same lollipop factory has eight workers to produce the 2968 lollipops. How many lollipops does each worker make in one week if they all work at the same rate?

20. A group of nine bricklayers win £4743 on the football pools. If they share the money equally, how much does each one receive?

10 Long multiplication

Examples

1.
```
   456
 X  14
 ─────
  4560
  1824
 ─────
  6384
```
Answer = 6384

2.
```
   324
 X  31
 ─────
  9720
   324
 ─────
 10 044
```
Answer = 10 044

Exercise 1

1a.
```
   312
 X  13
```
b.
```
   410
 X  22
```
c.
```
   134
 X  25
```
d.
```
   213
 X  37
```

2a.
```
   416
 X  51
```
b.
```
   394
 X  61
```
c.
```
   437
 X  32
```
d.
```
   536
 X  47
```

3a.
```
   431
 X  82
```
b.
```
   661
 X  27
```
c.
```
   422
 X  58
```
d.
```
   563
 X  23
```

4a.
```
   473
 X  73
```
b.
```
   569
 X  31
```
c.
```
   437
 X  94
```
d.
```
   458
 X  31
```

5a.
```
   536
 X  93
```
b.
```
   391
 X  36
```
c.
```
   403
 X  26
```
d.
```
   302
 X  48
```

6a.
```
  2435
 X   18
```
b.
```
  1748
 X   21
```
c.
```
  3164
 X   34
```
d.
```
  8175
 X   55
```

7. A pools group of fifteen workers shared out their win into equal amounts. If each person received £6945, how much was the total win?

8. 'Speedy Sports Cars' during one month sold 28 cars of the same model at £7850 each. How much money did they make altogether?

9. Binchester City had an average home crowd of 4862 supporters. If they played 24 home games in a season, find the total attendance for the year.

10. A newspaper works printed 4560 papers each day. If they sold all the copies on 25 consecutive days, how many were sold altogether?

11. An electrical shop sold 35 television sets costing £425 each. Find the total selling price.

12. 44 children paid £135 each for a holiday in Germany. How much is this altogether?

13. 52 workers in a factory turned out 3240 ballbearings each in a week. Find the total output.

14. 24 farm workers built a dry stone wall using 864 stones each. How many stones altogether?

15. 31 invaders from space each dropped 3250 leaflets on Earth telling it to surrender. How many leaflets did they need to print?

11 Fractions

Remember

A fraction is usually "part of a whole". Before you find any fraction, you must make sure that the whole is divided into *equal* parts.

Examples

1. Look at the shape.
    Answer these questions:

    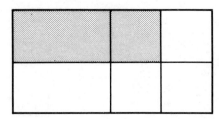

    How many equal parts is the whole circle divided into?   4

    How many parts are shaded?   1

    What fraction is shaded ? $\frac{1}{4}$

    What fraction is unshaded ? $\frac{3}{4}$

2. Look at the shape.
    Answer the question:

    How many equal parts altogether?

    First divide the shape into equal parts:
    There are **8** equal parts.

    How many parts are shaded?   3

    What fraction is shaded? $\frac{3}{8}$

    What fraction is unshaded? $\frac{5}{8}$

Exercise 1

Copy and complete the table. Make sure that each whole is divided into equal parts.

Question No.	1.	2.	3.	4.	5.	6.	7.	8.	9.	10.	11.	12.	13.	14.	15.
How many equal parts?															
How many parts are shaded?															
What fraction is shaded?															
What fraction is unshaded?															

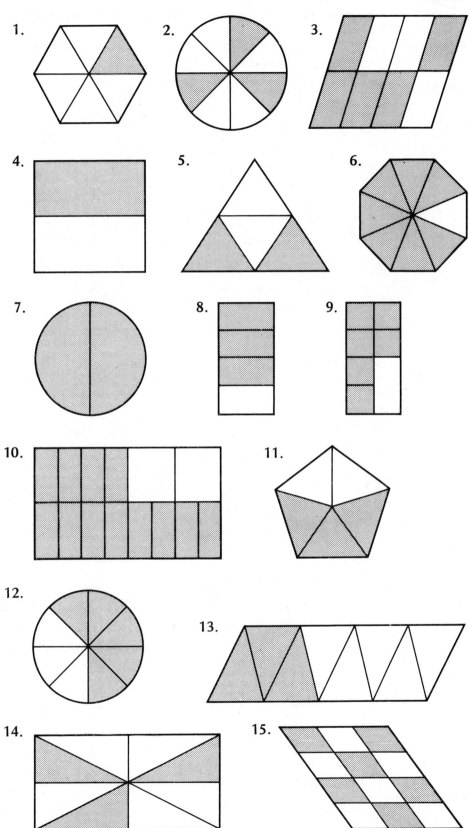

12 Shading in fractions

Remember

If you are given a fraction, then it is possible to shade in that fraction of a shape:

Examples

1. Copy and shade $\frac{2}{3}$ of this circle:

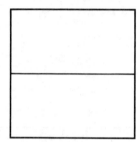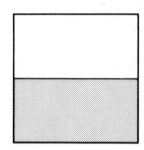

Sometimes you may have to divide the shape into equal parts yourself:

2. Copy and shade $\frac{1}{2}$ of this square: First, divide the square into *2 equal parts:*   Then shade *1* part:

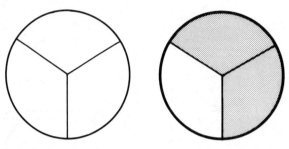

3. Copy and shade $\frac{5}{6}$ of this shape: First divide into *6 equal parts:*   Then shade 5 parts:

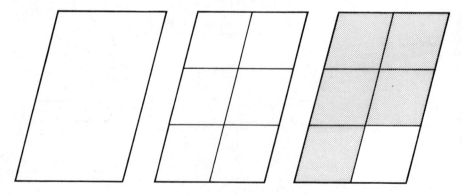

Exercise 1   Copy each shape and then shade the given fraction:

1. 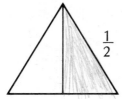 $\frac{1}{2}$

2.  $\frac{3}{4}$

3. 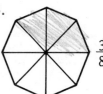 $\frac{3}{8}$

4. 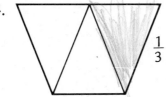 $\frac{1}{3}$

5.  $\frac{1}{4}$

6. 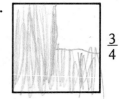 $\frac{3}{4}$

7.  $\frac{1}{2}$

8.  $\frac{1}{3}$

9.  $\frac{2}{3}$

10. 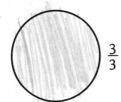 $\frac{3}{3}$

11. 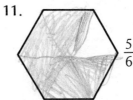 $\frac{5}{6}$

12.  $\frac{3}{5}$

13.  $\frac{7}{12}$

14. 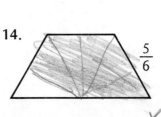 $\frac{5}{6}$

15. 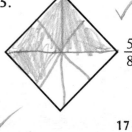 $\frac{5}{8}$

13 Equivalent fractions

Example

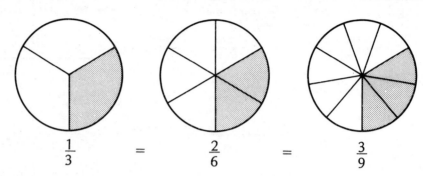

$$\frac{1}{3} \quad = \quad \frac{2}{6} \quad = \quad \frac{3}{9}$$

These fractions are equal.
They are called *equivalent fractions*.

Exercise 1   Find the equivalent fractions:

1.

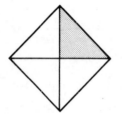  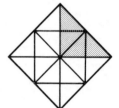

2.

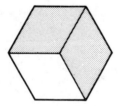 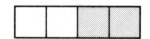 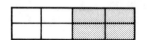

3.

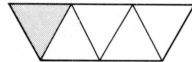 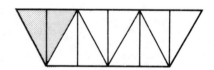

4.

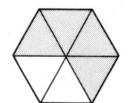 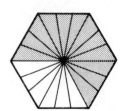

5.

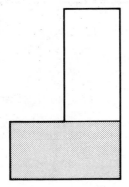 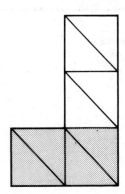

6.

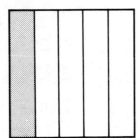 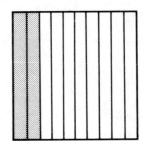 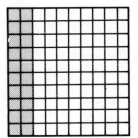

14 Finding equivalent fractions

Examples

1. Find 3 fractions that are equivalent to $\frac{1}{2}$.

$$\frac{1}{2} = \frac{1 \times 2}{2 \times 2} = \frac{2}{4}, \quad \frac{1}{2} = \frac{1 \times 3}{2 \times 3} = \frac{3}{6}, \quad \frac{1}{2} = \frac{1 \times 4}{2 \times 4} = \frac{4}{8}$$

so, $\frac{1}{2} = \frac{2}{4} = \frac{3}{6} = \frac{4}{8}$ These fractions are *equivalent*.

2. Find 3 fractions that are equivalent to $\frac{2}{3}$.

$$\frac{2}{3} = \frac{2 \times 2}{3 \times 2} = \frac{4}{6}, \quad \frac{2}{3} = \frac{2 \times 3}{3 \times 3} = \frac{6}{9}, \quad \frac{2}{3} = \frac{2 \times 4}{3 \times 4} = \frac{8}{12}$$

so, $\frac{2}{3} = \frac{4}{6} = \frac{6}{9} = \frac{8}{12}$ These fractions are *equivalent*.

Exercise 1

For each fraction, find 3 equivalent fractions:

1a. $\frac{1}{2}$ **b.** $\frac{1}{3}$ **c.** $\frac{1}{4}$ **d.** $\frac{1}{5}$ **e.** $\frac{1}{6}$

2a. $\frac{1}{8}$ **b.** $\frac{1}{10}$ **c.** $\frac{2}{3}$ **d.** $\frac{3}{4}$ **e.** $\frac{2}{5}$

3a. $\frac{3}{5}$ **b.** $\frac{4}{5}$ **c.** $\frac{5}{6}$ **d.** $\frac{3}{8}$ **e.** $\frac{7}{8}$

4a. $\frac{3}{10}$ **b.** $\frac{7}{10}$ **c.** $\frac{9}{10}$ **d.** $\frac{5}{12}$ **e.** $\frac{11}{12}$

15 Completing equivalent fractions

Examples

1. Complete: $\frac{3}{4} = \frac{*}{8}$ $\frac{3}{4} = \frac{6}{8}$ because 8 is twice 4, then * is twice 3, i.e. **6**.

2. Complete: $\frac{18}{20} = \frac{9}{*}$ $\frac{18}{20} = \frac{9}{10}$ because 9 is half of 18, then * is half of 20, i.e. **10**.

Exercise 1

Copy and complete the following equivalent fractions:

1a. $\frac{1}{2} = \frac{*}{4}$ b. $\frac{1}{2} = \frac{*}{8}$ c. $\frac{2}{3} = \frac{*}{6}$ d. $\frac{7}{10} = \frac{*}{30}$

2a. $\frac{2}{5} = \frac{*}{10}$ b. $\frac{2}{4} = \frac{4}{*}$ c. $\frac{4}{6} = \frac{12}{*}$ d. $\frac{2}{2} = \frac{6}{*}$

3a. $\frac{1}{4} = \frac{4}{*}$ b. $\frac{3}{8} = \frac{6}{*}$ c. $\frac{3}{5} = \frac{9}{*}$ d. $\frac{3}{3} = \frac{6}{*}$

4a. $\frac{2}{3} = \frac{20}{*}$ b. $\frac{4}{6} = \frac{12}{*}$ c. $\frac{3}{10} = \frac{6}{*}$ d. $\frac{3}{10} = \frac{*}{30}$

5a. $\frac{1}{100} = \frac{2}{*}$ b. $\frac{50}{100} = \frac{*}{300}$ c. $\frac{3}{100} = \frac{*}{200}$ d. $\frac{10}{100} = \frac{50}{*}$

6a. $\frac{8}{10} = \frac{4}{*}$ b. $\frac{14}{20} = \frac{7}{*}$ c. $\frac{20}{30} = \frac{4}{*}$ d. $\frac{4}{12} = \frac{*}{3}$

7a. $\frac{10}{15} = \frac{2}{*}$ b. $\frac{16}{20} = \frac{4}{*}$ c. $\frac{27}{30} = \frac{9}{*}$ d. $\frac{30}{40} = \frac{3}{*}$

8a. $\frac{*}{10} = \frac{3}{5}$ b. $\frac{*}{6} = \frac{2}{3}$ c. $\frac{36}{*} = \frac{12}{18}$ d. $\frac{27}{*} = \frac{9}{15}$

9a. $\frac{*}{15} = \frac{2}{5}$ b. $\frac{15}{*} = \frac{3}{4}$ c. $\frac{15}{25} = \frac{*}{5}$ d. $\frac{49}{56} = \frac{7}{*}$

10a. $\frac{9}{*} = \frac{3}{8}$ b. $\frac{*}{30} = \frac{5}{6}$ c. $\frac{15}{50} = \frac{*}{10}$ d. $\frac{32}{40} = \frac{4}{*}$

11a. $\frac{*}{3} = \frac{8}{12}$ b. $\frac{10}{*} = \frac{1}{3}$ c. $\frac{15}{21} = \frac{*}{7}$ d. $\frac{32}{44} = \frac{8}{*}$

12a. $\frac{8}{36} = \frac{*}{9}$ b. $\frac{24}{42} = \frac{4}{*}$ c. $\frac{*}{5} = \frac{6}{30}$ d. $\frac{9}{*} = \frac{27}{36}$

13a. $\frac{10}{25} = \frac{*}{5}$ b. $\frac{40}{45} = \frac{8}{*}$ c. $\frac{14}{21} = \frac{*}{3}$ d. $\frac{36}{42} = \frac{6}{*}$

14a. $\frac{10}{30} = \frac{*}{3}$ b. $\frac{6}{*} = \frac{42}{70}$ c. $\frac{11}{*} = \frac{55}{60}$ d. $\frac{36}{45} = \frac{12}{*}$

15a. $\frac{10}{*} = \frac{20}{26}$ b. $\frac{*}{9} = \frac{4}{36}$ c. $\frac{18}{54} = \frac{*}{6}$ d. $\frac{*}{3} = \frac{12}{36}$

16a. $\frac{40}{56} = \frac{10}{*}$ b. $\frac{12}{*} = \frac{24}{30}$ c. $\frac{6}{*} = \frac{48}{72}$ d. $\frac{15}{60} = \frac{*}{12}$

16 Cancelling down equivalent fractions

Remember

$$\frac{1}{2} = \frac{2}{4} = \frac{4}{8} = \frac{8}{16} = \frac{16}{32} = \frac{32}{64}$$

All of the fractions above are equivalent, but $\frac{1}{2}$ is in its *lowest terms*.
All the other fractions will *cancel down* to $\frac{1}{2}$. You should always cancel a fraction down to its lowest terms. You do this by dividing both top and bottom numbers by the largest number you can find.

Exercise 1

What is the largest number that you can find to divide both numbers in the following pairs (without leaving a remainder)?

1a. 4 and 6	b. 2 and 4	c. 4 and 8	d. 6 and 3
2a. 9 and 6	b. 10 and 5	c. 10 and 15	d. 30 and 100
3a. 8 and 12	b. 12 and 6	c. 21 and 14	d. 16 and 24
4a. 12 and 18	b. 20 and 16	c. 35 and 45	d. 40 and 50
5a. 10 and 100	b. 100 and 50	c. 100 and 200	d. 17 and 7

Examples

1. Cancel down $\frac{9}{12}$ to its lowest terms:

    $\frac{9}{12} = \frac{3}{4}$, the top and the bottom of the fraction have been divided by 3.

2. Cancel down $\frac{18}{24}$ to its lowest terms:

    $\frac{18}{24} = \frac{6}{8}$ (top and bottom divided by 3, but this can *still* be divided. . .)

    $\frac{6}{8} = \frac{3}{4}$ (top and bottom divided by 2)

    $\frac{3}{4}$ is in its *lowest terms* because there is no number you can use to divide *both* the 3 and 4.

Exercise 2

Cancel down each of the following fractions to its lowest terms:

1a. $\frac{2}{4}$ $\frac{1}{2}$ ✓	b. $\frac{3}{6}$ $\frac{1}{2}$ ✓	c. $\frac{4}{8}$ $\frac{2}{4}$	d. $\frac{5}{10}$ $\frac{1}{2}$ ✓
2a. $\frac{10}{20}$ $\frac{5}{10}$	b. $\frac{10}{30}$ $\frac{5}{15}$	c. $\frac{6}{8}$ $\frac{3}{4}$ ✓	d. $\frac{5}{15}$ $\frac{1}{3}$ ✓
3a. $\frac{8}{12}$ $\frac{4}{6}$	b. $\frac{14}{21}$ $\frac{2}{3}$ ✓	c. $\frac{10}{100}$ $\frac{5}{50}$ ✗	d. $\frac{100}{200}$ $\frac{50}{100}$
4a. $\frac{12}{20}$ $\frac{6}{10}$	b. $\frac{16}{28}$ $\frac{8}{14}$	c. $\frac{25}{30}$ $\frac{5}{6}$ ✓	d. $\frac{35}{40}$ $\frac{7}{8}$ ✓
5a. $\frac{24}{32}$ $\frac{12}{16}$	b. $\frac{16}{40}$ $\frac{8}{20}$	c. $\frac{100}{1000}$ $\frac{50}{500}$	d. $\frac{500}{1000}$ $\frac{250}{500}$
6a. $\frac{12}{30}$ $\frac{6}{15}$	b. $\frac{36}{48}$ $\frac{18}{24}$	c. $\frac{15}{35}$ $\frac{3}{7}$ ✓	d. $\frac{30}{45}$ $\frac{6}{9}$
7a. $\frac{24}{40}$ $\frac{12}{20}$	b. $\frac{32}{56}$ $\frac{16}{28}$	c. $\frac{18}{36}$ $\frac{9}{18}$	d. $\frac{27}{45}$ $\frac{9}{15}$
8a. $\frac{33}{55}$ $\frac{3}{5}$ ✓	b. $\frac{44}{99}$ $\frac{4}{9}$ ✓	c. $\frac{24}{60}$ $\frac{12}{20}$	d. $\frac{48}{72}$ $\frac{24}{36}$

17 Mixed numbers

Remember

A *mixed number* is a mixture of a whole number and a fraction, e.g.

$1\frac{1}{2}$ *is a mixed number*

An *improper fraction* has a bigger number on the top than on the bottom, e.g.

$\frac{5}{4}$ *is an improper fraction*

Improper fractions can be changed to mixed numbers. You can draw a picture to show this:

Examples

1. Change $\frac{5}{4}$ to a mixed number.

   $\frac{5}{4}$ is 5 quarters:

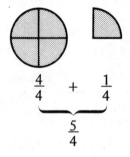

$$\frac{4}{4} \quad + \quad \frac{1}{4}$$

$$\frac{5}{4}$$

See that $\frac{5}{4} = 1\frac{1}{4}$

2. Change $\frac{11}{4}$ to a mixed number.

   $\frac{11}{4}$ is 11 quarters. As a picture:

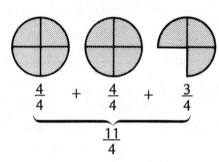

$$\frac{4}{4} \quad + \quad \frac{4}{4} \quad + \quad \frac{3}{4}$$

$$\frac{11}{4}$$

See that $\frac{11}{4} = 2\frac{3}{4}$

Exercise 1

1. Write down 5 examples of a mixed number.
2. Write down 5 examples of numbers which are *not* mixed.
3. Write down 5 examples of an improper fraction.
4. Write down 5 examples of fractions which are *not* improper.

Exercise 2

Change each improper fraction to a mixed number:

1a. $\frac{4}{3}$ b. $\frac{5}{3}$ c. $\frac{11}{8}$ d. $\frac{15}{8}$ e. $\frac{13}{6}$

2a. $\frac{7}{6}$ b. $\frac{11}{6}$ c. $\frac{7}{5}$ d. $\frac{13}{10}$ e. $\frac{14}{5}$

3a. $\frac{9}{2}$ b. $\frac{7}{3}$ c. $\frac{8}{3}$ d. $\frac{11}{4}$ e. $\frac{17}{3}$

4a. $\frac{11}{2}$ b. $\frac{11}{3}$ c. $\frac{13}{4}$ d. $\frac{10}{3}$ e. $\frac{8}{5}$

5a. $\frac{12}{5}$ b. $\frac{23}{10}$ c. $\frac{13}{5}$ d. $\frac{5}{2}$ e. $\frac{19}{2}$

Remember

It is possible to cancel down a fraction to its lowest terms, and then change it to a mixed number:

Examples

1. $\frac{10}{4} = \frac{5}{2} = 2\frac{1}{2}$ 2. $\frac{10}{8} = \frac{5}{4} = 1\frac{1}{4}$ 3. $\frac{12}{9} = \frac{4}{3} = 1\frac{1}{3}$

Exercise 3   1a. $\frac{6}{4}$ b. $\frac{9}{6}$ c. $\frac{12}{8}$ d. $\frac{22}{10}$

2a. $\frac{14}{8}$ b. $\frac{25}{10}$ c. $\frac{8}{6}$ d. $\frac{15}{10}$

3a. $\frac{10}{6}$ b. $\frac{15}{6}$ c. $\frac{150}{100}$ d. $\frac{18}{8}$

18 Using mixed numbers

Remember

Understanding fractions gives you another way of dealing with a division that leaves a remainder:

Examples

1. Find $43 \div 4$. $4\overline{)43}$ = 10 r. 3   So $43 \div 4 = 10\frac{3}{4}$

2. Find $17 \div 2$. $2\overline{)17}$ = 8 r. 1   So $17 \div 2 = 8\frac{1}{2}$

3. Find $25 \div 10$. $10\overline{)25}$ = 2 r. 5   So $25 \div 10 = 2\frac{5}{10} = 2\frac{1}{2}$

Exercise 1

1a. $9 \div 2$ b. $9 \div 4$ c. $19 \div 4$ d. $27 \div 10$

2a. $13 \div 4$ b. $13 \div 5$ c. $15 \div 4$ d. $15 \div 10$

3a. $26 \div 5$ b. $31 \div 8$ c. $27 \div 8$ d. $19 \div 3$

4a. $17 \div 3$ b. $28 \div 4$ c. $18 \div 2$ d. $36 \div 8$

5a. $53 \div 10$ b. $76 \div 10$ c. $39 \div 10$ d. $131 \div 10$

In fact, a fraction is exactly the same as the division of one number by another:

Examples

1. $9 \div 2$ is the same as $\frac{9}{2}$.

2. $5 \div 4$ is the same as $\frac{5}{4}$.

3. Rewrite $5 \div 3$ as a fraction.

   $5 \div 3 = \frac{5}{3}$

Exercise 2

Rewrite each division as a fraction and cancel down where possible:

1a. $7 \div 4$ b. $24 \div 5$ c. $13 \div 8$ d. $19 \div 4$

2a. $14 \div 4$ b. $4 \div 6$ c. $6 \div 12$ d. $9 \div 6$

3a. $5 \div 2$ b. $7 \div 3$ c. $9 \div 4$ d. $11 \div 5$

4a. $3 \div 7$ b. $4 \div 5$ c. $9 \div 10$ d. $6 \div 8$

5a. $4 \div 9$ b. $11 \div 3$ c. $13 \div 5$ d. $5 \div 9$

19 Metric units

Remember   The Metric System is easy to use. All the units work in multiples of 10:

length has basic unit the **metre**:

10 millimetres (mm)	= 1 centimetre (cm)
100 centimetres (cm)	= 1 metre (m)
1000 metres (m)	= 1 kilometre (km)

mass has basic unit the **gram**:

1000 grams	(g) = 1 kilogram (kg)
1000 kilograms	(kg) = 1 tonne (t)

volume has basic unit the **litre**:

1000 millilitres   (ml) = 1 litre (l)

Example   1 cm = * mm   **1 cm = 10 mm**

Exercise 1   Copy and complete the following:

1a. 10 mm = * cm	b. 30 mm = * cm	c. 60 mm = * cm
2a. 100 cm = * m	b. 300 cm = * m	c. 700 cm = * m
3a. 1000 m = * km	b. 4000 m = * km	c. 8000 m = * km
4a. 1000 g = * kg	b. 5000 g = * kg	c. 9000 g = * kg
5a. 1000 kg = * t	b. 6000 kg = * t	c. 8000 kg = * t
6a. 1000 ml = * l	b. 4000 ml = * l	c. 7000 ml = * l
7a. * mm = 6 cm	b. * mm = 12 cm	c. * mm = 20 cm
8a. * m = 200 cm	b. * m = 500 cm	c. * m = 900 cm
9a. * km = 3000 m	b. * km = 5000 m	c. * km = 9000 m
10a. * kg = 2000 g	b. * kg = 3000 g	c. * kg = 7000 g

Exercise 2

1. One of the recognised Olympic track events is the 10 000 m race. How many kilometres is this?

2. A metre stick is exactly a metre long. If you put 2000 metre sticks end to end in a straight line, how far would they reach in kilometres?

3. If you were given a balance with a $\frac{1}{4}$ kilogram weight in one scale-pan, how many 1 gram weights would you need to put in the other pan for them to balance?

4. Ten bottles each contain 200 ml when full. If they were all emptied into the same container, how many ml must this container be able to hold?

5. Angabout the tortoise has moved 10 cm during the time that you have walked 800 cm.
   a. How many millimetres has Angabout moved?
   b. How many metres have you walked?

20 Decimal numbers

Remember

A decimal point separates a number into its *whole number part* and its *fractional part*:

Examples

1. 23·4 has whole number part **23** and fractional part $\frac{4}{10}$

2. 3·24 has whole number part **3** and fractional parts $\frac{2}{10}$ and $\frac{4}{100}$

Exercise 1

Copy and complete the table below showing the whole number and fractional parts of each decimal number given. The first one has been done for you:

	whole number	fractional parts		
		tenths	hundredths	thousandths
24·402	24	$\frac{4}{10}$	$\frac{0}{100}$	$\frac{2}{1000}$
18·213				
7·4				
6·811				
1892·002				

Exercise 2

Copy and complete the following table by placing the figures of each decimal number in the correct spaces. The first one has been done for you:

	1000	100	10	1	$\frac{1}{10}$	$\frac{1}{100}$	$\frac{1}{1000}$
34·836			3	4	8	3	6
3·806							
39·999							
0·583							
345·76							
498·3							
31·007							
1567·92							
1286·44							
2316·804							

21 Using decimal numbers

Examples

1. 24·378 is **twenty-four point three, seven, eight**

2. 104·302 is **one hundred and four point three, zero, two**

Exercise 1

Write each of the following decimal numbers in words:

1a. 4·2	**b.** 18·6	**c.** 2·42	**d.** 1·61
2a. 10·88	**b.** 39·01	**c.** 50·202	**d.** 148·3
3a. 2·399	**b.** 45·890	**c.** 2348·1	**d.** 3·549

Exercise 2

Write each written decimal *in figures*:

1. Forty-two point six.

2. Eighteen point seven, one, two.

3. Four hundred and twelve point nine, three.

4. Nine point eight, zero, five.

5. Fifty-seven point zero, one, seven.

6. One thousand and six point six, two.

7. Zero point four, four, four.

8. Ten point nine, zero, nine.

9. Eight hundred and four point zero, six, eight.

10. Six thousand, three hundred and thirty-four point zero, one.

22 Comparing decimal numbers

Remember

To compare the size of two decimal numbers, start with the *whole number* parts. If one whole number part is larger than the other, then that is the larger decimal number, *no matter what fractional parts there are.*

Examples

Arrange these decimal numbers in order, with the largest number written first:

1. 21·49 and 18·949
   21·49, 18·949   (21 is larger than 18)

2. 7·4, 12·03 and 8·969
   12·03, 8·969, 7·4   (12 is larger than 8 which is larger than 7)

Exercise 1   Arrange the following decimal numbers in order, largest first:

1a. 3·4, 2·8   b. 7·49, 9·21

2a. 1·4, 2·1   b. 14·99, 12·999

3a. 17·6, 77·1   b. 20·906, 19·999

4a. 2·3, 4·3, 3·2   b. 2·44, 3·11, 1·232

5a. 14·6, 144·6, 74·22   b. 129·5, 29·87, 4·736

Examples   Arrange each set of decimal numbers in order, with the largest number written first:

1. 21·49 and 21·54
    The whole number parts
    are the same, so compare the
    figures for the *tenths*:
    21·**4**9 and 21·**5**4
    5 tenths is larger than **4** tenths,
    so the order is:
    21·54, 21·49

2. 102·441 and 102·46
    The whole number parts
    and the tenths parts are the
    same, so compare the figures
    for the *hundredths*:
    102·4**4**1 and 102·4**6**
    6 hundredths is larger than
    4 hundredths, so the order is:
    102·46, 102·441

Exercise 2   Arrange the following decimal numbers in order, largest first:

1a. 3·6, 3·8   b. 14·6, 14·2

2a. 4·61, 4·70   b. 73·8, 73·6

3a. 38·4, 38·29   b. 3·901, 3·911

4a. 17·62, 17·606   b. 512·66, 512·861

5a. 112·63, 112·333   b. 90·009, 90·1

6a. 4·6, 4·2, 4·3   b. 10·09, 10·19, 10·18

7a. 7·44, 7·43, 7·46   b. 800·9, 800·85, 800·87

8a. 62·1, 62·18   b. 16·859, 16·85

9a. 3·446, 3·442, 3·443   b. 4·831, 4·83, 4·829

10a. 10·641, 10·644, 10·645   b. 1·723, 1·726, 1·725

11a. 3·86, 3·858, 3·856   b. 9·816, 9·814, 9·815

12a. 22·621, 22·624, 22·623   b. 91·066, 91·063, 91·06

13a. 5·332, 5·341, 5·337   b. 10·409, 10·40, 10·41

14a. 12·6, 12·60, 12·61   b. 8·324, 8·323, 8·32

15a. 4·66, 4·666, 4·6   b. 3·91, 3·903, 3·906

23 Changing decimal numbers to fractions

Remember   A decimal number can be written as a fraction very easily.

Examples   1. $0.5 = \frac{5}{10}$ 2. $0.1 = \frac{1}{10}$

        which cancels to $\frac{1}{2}$, so $\mathbf{0.5} = \frac{1}{2}$

       3. $0.25 = \frac{25}{100}$ 4. $0.003 = \frac{3}{1000}$

        which cancels to $\frac{1}{4}$, so $\mathbf{0.25} = \frac{1}{4}$

       You need to *remember* that $\mathbf{0.5} = \frac{1}{2}$ and that $\mathbf{0.25} = \frac{1}{4}$

Exercise 1   Write the following decimal numbers as fractions but then cancel them down to their lowest terms, when necessary:

1a. 0·1	b. 0·3	c. 0·7	d. 0·9
2a. 0·2	b. 0·4	c. 0·6	d. 0·8
3a. 0·11	b. 0·17	c. 0·23	d. 0·29
4a. 0·14	b. 0·26	c. 0·34	d. 0·46
5a. 0·03	b. 0·07	c. 0·06	d. 0·08
6a. 0·007	b. 0·009	c. 0·004	d. 0·008
7a. 0·16	b. 0·32	c. 0·55	d. 0·85
8a. 0·013	b. 0·017	c. 0·024	d. 0·036
9a. 0·242	b. 0·468	c. 0·25	d. 0·75
10a. 0·125	b. 0·375	c. 0·625	d. 0·875

24 Changing fractions to decimal numbers

Remember   A fraction can also be written as a decimal number:

Examples   1. $\frac{3}{10} = 0.3$ 2. $\frac{18}{100} = 0.18$ 3. $\frac{7}{100} = 0.07$

       4. $\frac{24}{1000} = 0.024$ 5. $\frac{9}{1000} = 0.009$

Exercise 1   Write the following fractions as decimal numbers:

1a. $\frac{1}{10}$	b. $\frac{3}{10}$	c. $\frac{5}{10}$	d. $\frac{8}{10}$
2a. $\frac{1}{100}$	b. $\frac{4}{100}$	c. $\frac{7}{100}$	d. $\frac{9}{100}$
3a. $\frac{14}{100}$	b. $\frac{36}{100}$	c. $\frac{45}{100}$	d. $\frac{69}{100}$
4a. $\frac{1}{1000}$	b. $\frac{8}{1000}$	c. $\frac{11}{1000}$	d. $\frac{24}{1000}$
5a. $\frac{112}{1000}$	b. $\frac{235}{1000}$	c. $\frac{486}{1000}$	d. $\frac{888}{1000}$

Remember   If a fraction does not have a denominator (bottom number) of 10 or 100 or 1000 or . . . . then try to find an equivalent fraction that does. Then write that as a decimal number:

Examples

1. $\frac{1}{2} = \frac{5}{10} = 0.5$ so, $\frac{1}{2} = \mathbf{0.5}$ 2. $\frac{3}{5} = \frac{6}{10} = \mathbf{0.6}$

3. $\frac{9}{50} = \frac{18}{100} = 0.18$ 4. $\frac{1}{20} = \frac{5}{100} = 0.05$ 5. $\frac{7}{200} = \frac{35}{1000} = 0.035$

Exercise 2

1a. $\frac{1}{5}$ b. $\frac{2}{5}$ c. $\frac{3}{20}$ d. $\frac{3}{50}$ e. $\frac{4}{5}$

2a. $\frac{7}{20}$ b. $\frac{7}{50}$ c. $\frac{11}{50}$ d. $\frac{11}{20}$ e. $\frac{21}{50}$

3a. $\frac{4}{20}$ b. $\frac{49}{50}$ c. $\frac{1}{4}$ d. $\frac{2}{4}$ e. $\frac{3}{4}$

4a. $\frac{15}{20}$ b. $\frac{101}{500}$ c. $\frac{200}{500}$ d. $\frac{203}{500}$ e. $\frac{408}{500}$

5a. $\frac{3}{15}$ b. $\frac{6}{15}$ c. $\frac{6}{25}$ d. $\frac{13}{25}$ e. $\frac{21}{25}$

Remember   Mixed numbers are easy to convert to decimal numbers. The whole number part of the mixed number is put before the decimal point; the fractional part is dealt with as above.

1. Write $3\frac{1}{2}$ as a decimal number.   2. Write $2\frac{3}{10}$ as a decimal number.

$\frac{1}{2} = 0.5$, so $3\frac{1}{2} = \mathbf{3.5}$ $\frac{3}{10} = 0.3$, so $2\frac{3}{10} = \mathbf{2.3}$

3. Write $5\frac{1}{25}$ as a decimal number.   4. Write $6\frac{19}{250}$ as a decimal number.

$\frac{1}{25} = \frac{1 \times 4}{100} = \frac{4}{100} = 0.04.$ $\frac{19}{250} = \frac{19 \times 4}{1000} = \frac{76}{1000} = 0.076.$

So $5\frac{1}{25} = \mathbf{5.04}$ So $6\frac{19}{250} = \mathbf{6.076}$

Exercise 3   Write the following mixed numbers as decimal numbers:

1a. $1\frac{1}{10}$ b. $2\frac{3}{10}$ c. $3\frac{6}{10}$ d. $4\frac{9}{10}$ e. $5\frac{3}{100}$

2a. $6\frac{27}{100}$ b. $7\frac{23}{100}$ c. $8\frac{44}{100}$ d. $1\frac{4}{1000}$ e. $2\frac{9}{1000}$

3a. $3\frac{18}{1000}$ b. $4\frac{58}{1000}$ c. $5\frac{124}{1000}$ d. $6\frac{304}{1000}$ e. $7\frac{466}{1000}$

4a. $8\frac{875}{1000}$ b. $8\frac{1}{5}$ c. $3\frac{2}{5}$ d. $4\frac{3}{5}$ e. $6\frac{4}{5}$

5a. $2\frac{3}{20}$ b. $5\frac{7}{20}$ c. $6\frac{13}{20}$ d. $3\frac{16}{20}$ e. $4\frac{14}{20}$

6a. $1\frac{26}{50}$ b. $7\frac{38}{50}$ c. $6\frac{41}{50}$ d. $3\frac{31}{50}$ e. $8\frac{47}{50}$

7a. $7\frac{121}{200}$ b. $5\frac{166}{200}$ c. $4\frac{126}{200}$ d. $5\frac{165}{200}$ e. $4\frac{187}{200}$

8a. $4\frac{27}{250}$ b. $6\frac{81}{250}$ c. $3\frac{123}{250}$ d. $2\frac{166}{250}$ e. $7\frac{207}{250}$

9a. $8\frac{87}{500}$ b. $5\frac{129}{500}$ c. $9\frac{212}{500}$ d. $6\frac{365}{500}$ e. $5\frac{419}{500}$

10a. $2\frac{8}{25}$ b. $7\frac{14}{25}$ c. $8\frac{21}{25}$ d. $5\frac{1}{4}$ e. $6\frac{3}{4}$

25 Decimal places

The number of figures after the decimal point gives the number of decimal places in a decimal number. Zeros are not counted if they occur *at the end* of the decimal number.

Examples
1. 1062·604 **has 3 decimal places** (the zero in the middle is counted)
2. 1·1630 **has 3 decimal places** (the zero at the end is not counted)

Exercise 1

How many decimal places have each of the following decimal numbers?

1a.	0·4	b.	0·8	c.	0·9
2a.	0·15	b.	0·36	c.	0·48
3a.	0·324	b.	0·675	c.	0·824
4a.	0·05	b.	0·03	c.	0·07
5a.	0·009	b.	0·013	c.	0·099
6a.	2·308	b.	3·504	c.	4·607
7a.	12·620	b.	18·704	c.	26·730
8a.	31·799	b.	46·800	c.	51·839
9a.	96·003	b.	100·007	c.	162·009

Remember

A number often needs to be written to a given number of decimal places. To decide upon whether to 'round down' or 'round up' you look at the *next, smaller* decimal place, which is one decimal place to the *right*. If it is *less* than 5 you round down; if it is 5 or more, you round up.

Examples
1. Write 25·4723 to 2 d.p.
   Look at the *3rd* decimal place:
   The 3rd place is a '2' so you round down:
   25·4723 to 2 decimal places = 25·47

2. Write 14·76 to 1 d.p.
   Look at the *2nd* decimal place.
   The 2nd place is a '6' so you round up:
   14·76 to 1 decimal place = 14·8

3. Write 14·754 to 1 d.p.
   Look at the *2nd* decimal place.
   The 2nd place is a '5' so you round up:
   14·754 to 1 decimal place = 14·8

Exercise 2

Write each decimal number to the given number of places:

1a.	1·64	to 1 d.p.	b.	8·63	to 1 d.p.
2a.	4·863	to 2 d.p.	b.	5·652	to 2 d.p.
3a.	9·3264	to 3 d.p.	b.	9·9833	to 3 d.p.

4a. 8·6432 to 2 d.p.   b. 6·4626 to 2 d.p.

5a. 3·47 to 1 d.p.   b. 8·69 to 1 d.p.

6a. 4·346 to 2 d.p.   b. 7·465 to 2 d.p.

7a. 8·0686 to 3 d.p.   b. 5·4038 to 3 d.p.

8a. 6·2963 to 2 d.p.   b. 4·3068 to 2 d.p.

9a. 8·326 to 1 d.p.   b. 6·492 to 2 d.p.

10a. 12·622 to 2 d.p.   b. 14·4326 to 1 d.p.

26 Significant figures

Remember

To write a number to a given number of *significant figures* is exactly the same as writing a number to a given number of decimal places except that you must count *all* the figures in the number, not just the decimal part:

Examples

1. Write 13·628 to 4 significant figures.
   The 5th figure is '8', so
   13·628 = 13·63 (to 4 s.f.)

2. Write 16·032 to 2 significant figures.
   The 3rd figure is '0', so
   16·032 = 16 (to 2 s.f.)

3. Write 0·4892 to 2 significant figures.
   The 3rd figure is '9', (don't count any zero at the left of the number),
   so,   **0·4892 = 0·49 (to 2 s.f.)**

Exercise 1

Write each number to the given number of significant figures:

1a. 1·324 to 3 s.f.   b. 8·615 to 2 s.f.

2a. 12·86 to 3 s.f.   b. 49·47 to 3 s.f.

3a. 156·26 to 4 s.f.   b. 7·266 to 3 s.f.

4a. 7·266 to 2 s.f.   b. 7·266 to 1 s.f.

5a. 4·85 to 2 s.f.   b. 4·851 to 2 s.f.

6a. 0·726 to 2 s.f.   b. 0·726 to 1 s.f.

7a. 0·048 to 1 s.f.   b. 0·0074 to 1 s.f.

8a. 4·007 to 1 s.f.   b. 4·007 to 2 s.f.

9a. 14·007 to 2 s.f.   b. 14·851 to 3 s.f.

10a. 3·95 to 2 s.f.   b. 13·999 to 2 s.f.

27 Decimals: true or false?

Are the following statements true or false?

1. The 4 in 5·342 stands for 4 hundredths.

2. 21·7 is smaller than 21·66.

3. 52·10 is the same as 52·1.

4. ·62 is equal to 0·62.

5. The 6 in 26·325 stands for six tenths.

6. The 8 in 6·448 stands for eight thousandths.

7. 3·59 is larger than 3·7.

8. 0·04 is smaller than 0·004.

9. 0·5 is the same as $\frac{1}{2}$.

10. $\frac{1}{4}$ is the same as 0·25.

11. $\frac{1}{8}$ is the same as 0·125.

12. 40·3 is smaller than 40·28.

13. The 6 in 46·42 stands for six tens.

14. $\frac{3}{4}$ is the same as 0·75.

15. 0·625 is the same as $\frac{5}{8}$.

16. 24·44 is written to two decimal places.

17. 10·62 is written to three significant figures.

18. 0·0043 is written to two decimal places.

19. 0·0044 is written to two significant figures.

20. 60·04 is written to two significant figures.

21. 23·0 has the same value as 230.

22. 230 has the same value as 230·0.

23. The decimal value of $\frac{1}{5}$ is 0·4.

24. 0·8 can be written as $\frac{8}{100}$.

25. 0·007 can be written as $\frac{7}{1000}$.

28 Decimal addition

Remember

Addition of decimal numbers is similar to addition of whole numbers. But you must make sure that you add the whole number parts and fractional parts separately:

Examples

1.
```
  6·4
+3·2
─────
  9·6
```
Notice how the decimal points are all in line to keep the whole number and fractional parts separate.

2.
```
  14·6
+  7·2
──────
  21·8
```

3. Find 14·37 + 4·82
```
  14·37
+  4·82
───────
  19·19
```

4. Add together 175·5 and 8·25
```
  175·5
+   8·25
────────
  183·75
```

Exercise 1

Be careful to keep the decimal points in a line as you do these examples.

1a.
```
  5·7
+2·2
```
b.
```
  3·6
+3·2
```
c.
```
  4·72
+3·17
```
d.
```
  5·101
+3·736
```

2a.
```
  3·8
+5·4
```
b.
```
  4·5
+1·7
```
c.
```
  3·67
+3·28
```
d.
```
  5·82
+2·66
```

3a.
```
  4·38
+3·43
```
b.
```
  14·19
+  3·73
```
c.
```
  1·16
+0·67
```
d.
```
  1·36
+0·58
```

4a.
```
  26·73
+  2·46
```
b.
```
  36·2
+10·8
```
c.
```
  5·64
+3·58
```
d.
```
  8·23
+0·79
```

5a.
```
  17·43
+  3·5
```
b.
```
  45·6
+  3·66
```
c.
```
  46·78
+17·2
```
d.
```
    4·88
+231·0
```

6a.
```
  4·6
  2·1
+1·5
```
b.
```
  6·7
  2·4
+4·6
```
c.
```
  23·4
  14·8
+14·1
```
d.
```
  4·30
  5·8
+3·19
```

7a.
```
   6·518
  12·572
+  3·7
```
b.
```
   31·46
   28·174
+130·1
```
c.
```
  0·449
  1·777
+0·11
```
d.
```
  4·64
  6·099
+3·71
```

8. A man who is 1·73 metres tall is able to jump a further 0·50 metres off the ground. If he were playing soccer what is the highest ball that he could reach with his head?

9. Add together three point four, nine; sixteen point eight; and forty-two point two, seven, one.

29 Decimal subtraction

Remember

Subtraction of decimal numbers is similar to subtraction of whole numbers. But you must keep the whole number and fractional parts separate:

Examples

1.	2·9	2.	15·6	3.	27·05
	−1·4		− 3·8		− 9·1
	1·5		**11·8**		**17·95**

4.	6·43	5.	219·65	6.	6·40
	−3·85		− 55·7		−4·57
	2·58		**163·95**		**1·83**

Exercise 1

Do these decimal subtractions:

1a.	5·8	b.	8·5	c.	9·8	d.	6·9
	−3·6		−5·3		−6·5		−1·4

2a.	4·76	b.	5·39	c.	7·66	d.	8·45
	−2·35		−1·17		−3·42		−6·13

3a.	6·4	b.	7·5	c.	8·4	d.	9·3
	−2·6		−4·6		−3·8		−6·8

4a.	6·36	b.	6·65	c.	3·27	d.	4·86
	−2·18		−3·37		−1·09		−2·38

5a.	9·54	b.	3·62	c.	4·43	d.	8·64
	−4·62		−1·71		−2·82		−5·71

6a.	6·34	b.	7·46	c.	2·65	d.	8·45
	−2·66		−5·58		−1·97		−4·98

7a.	6·324	b.	5·656	c.	6·374	d.	8·357
	−3·612		−2·834		−4·561		−5·745

8a.	8·236	b.	6·454	c.	6·365	d.	7·447
	−3·378		−3·666		−2·689		−4·579

9a.	8·68	b.	7·53	c.	9·68	d.	6·43
	−5·4		−3·3		−6·2		−4·1

10a.	7·32	b.	6·54	c.	5·61	d.	8·13
	−4·6		−3·7		−2·8		−3·9

Example

$$
\begin{array}{r}
7{\cdot}4 \\
-1{\cdot}37 \\
\hline
\end{array}
$$

the '7' has nothing to be subtracted from.

Rewrite the subtraction like this:

$$
\begin{array}{r}
7{\cdot}40 \\
-1{\cdot}37 \\
\hline
\mathbf{6{\cdot}03} \\
\hline
\end{array}
$$

which is worked in the same way as example 6 on the last page.

You can always add zeros at the *end* of a decimal number after the decimal point; the value of the decimal number remains the same.

Exercise 2

Do these decimal subtractions:

1a. $\begin{array}{r} 8{\cdot}6 \\ -6{\cdot}42 \\ \hline \end{array}$ b. $\begin{array}{r} 9{\cdot}3 \\ -7{\cdot}05 \\ \hline \end{array}$ c. $\begin{array}{r} 7{\cdot}4 \\ -3{\cdot}26 \\ \hline \end{array}$ d. $\begin{array}{r} 5{\cdot}6 \\ -2{\cdot}49 \\ \hline \end{array}$

2a. $\begin{array}{r} 6{\cdot}3 \\ -2{\cdot}53 \\ \hline \end{array}$ b. $\begin{array}{r} 4{\cdot}6 \\ -1{\cdot}84 \\ \hline \end{array}$ c. $\begin{array}{r} 5{\cdot}1 \\ -2{\cdot}52 \\ \hline \end{array}$ d. $\begin{array}{r} 8{\cdot}5 \\ -6{\cdot}83 \\ \hline \end{array}$

3a. $\begin{array}{r} 7{\cdot}55 \\ -2{\cdot}439 \\ \hline \end{array}$ b. $\begin{array}{r} 6{\cdot}36 \\ -3{\cdot}222 \\ \hline \end{array}$ c. $\begin{array}{r} 8{\cdot}46 \\ -5{\cdot}135 \\ \hline \end{array}$ d. $\begin{array}{r} 9{\cdot}35 \\ -4{\cdot}126 \\ \hline \end{array}$

4a. $\begin{array}{r} 5{\cdot}63 \\ -2{\cdot}266 \\ \hline \end{array}$ b. $\begin{array}{r} 8{\cdot}74 \\ -3{\cdot}376 \\ \hline \end{array}$ c. $\begin{array}{r} 6{\cdot}56 \\ -4{\cdot}278 \\ \hline \end{array}$ d. $\begin{array}{r} 4{\cdot}26 \\ -1{\cdot}085 \\ \hline \end{array}$

5a. $\begin{array}{r} 8{\cdot}64 \\ -2{\cdot}752 \\ \hline \end{array}$ b. $\begin{array}{r} 7{\cdot}23 \\ -1{\cdot}343 \\ \hline \end{array}$ c. $\begin{array}{r} 5{\cdot}17 \\ -4{\cdot}281 \\ \hline \end{array}$ d. $\begin{array}{r} 9{\cdot}33 \\ -6{\cdot}678 \\ \hline \end{array}$

6a. $\begin{array}{r} 94{\cdot}6 \\ -38{\cdot}45 \\ \hline \end{array}$ b. $\begin{array}{r} 85{\cdot}5 \\ -36{\cdot}16 \\ \hline \end{array}$ c. $\begin{array}{r} 41{\cdot}6 \\ -14{\cdot}33 \\ \hline \end{array}$ d. $\begin{array}{r} 72{\cdot}7 \\ -45{\cdot}24 \\ \hline \end{array}$

7a. $\begin{array}{r} 56{\cdot}5 \\ -27{\cdot}62 \\ \hline \end{array}$ b. $\begin{array}{r} 74{\cdot}6 \\ -35{\cdot}81 \\ \hline \end{array}$ c. $\begin{array}{r} 35{\cdot}2 \\ -17{\cdot}54 \\ \hline \end{array}$ d. $\begin{array}{r} 43{\cdot}4 \\ -25{\cdot}53 \\ \hline \end{array}$

8a. $\begin{array}{r} 465{\cdot}5 \\ -\ 83{\cdot}13 \\ \hline \end{array}$ b. $\begin{array}{r} 324{\cdot}6 \\ -\ 62{\cdot}24 \\ \hline \end{array}$ c. $\begin{array}{r} 518{\cdot}3 \\ -\ 53{\cdot}17 \\ \hline \end{array}$ d. $\begin{array}{r} 862{\cdot}5 \\ -\ 71{\cdot}23 \\ \hline \end{array}$

9a. $\begin{array}{r} 543{\cdot}7 \\ -\ 20{\cdot}83 \\ \hline \end{array}$ b. $\begin{array}{r} 465{\cdot}3 \\ -\ 33{\cdot}67 \\ \hline \end{array}$ c. $\begin{array}{r} 621{\cdot}2 \\ -\ 10{\cdot}73 \\ \hline \end{array}$ d. $\begin{array}{r} 754{\cdot}6 \\ -\ 21{\cdot}89 \\ \hline \end{array}$

10a. $\begin{array}{r} 642{\cdot}6 \\ -\ 56{\cdot}93 \\ \hline \end{array}$ b. $\begin{array}{r} 875{\cdot}4 \\ -\ 86{\cdot}77 \\ \hline \end{array}$ c. $\begin{array}{r} 736{\cdot}8 \\ -\ 68{\cdot}91 \\ \hline \end{array}$ d. $\begin{array}{r} 666{\cdot}6 \\ -\ 79{\cdot}84 \\ \hline \end{array}$

30 Multiplication of decimals by numbers less than 10

Examples

1. How much would three pairs of socks cost if one pair costs £1·32?
To get the answer multiply £1·32 by 3:

   £1·32
   X   3
   ────
   £3·96

   Notice that the decimal point has been written down in the answer underneath the decimal point in the question.

   Answer = £3·96

2.   2·42
    X   7
    ─────
    16·94

3. Find 3·52 X 4

      3·52
    X   4
    ─────
    14·08

Exercise 1

Do these decimal multiplications:

1a. 4·1
   X 2

b. 3·2
   X 3

c. 5·4
   X 3

d. 3·7
   X 3

2a. 5·6
   X 4

b. 12·4
   X   4

c. 10·22
   X   5

d. 103·33
   X   5

3a. 14·8
   X   7

b. 24·7
   X   5

c. 12·5
   X   3

d. 36·9
   X   4

4a. 2·55
   X   6

b. 6·34
   X   8

c. 4·83
   X   7

d. 9·32
   X   4

5a. 29·4
   X   5

b. 6·13
   X   7

c. 28·01
   X   5

d. 6·24
   X   9

6a. 2·14 X 5   b. 3·61 X 2   c. 4·09 X 3

7a. 4·59 X 2   b. 6·72 X 3   c. 5·61 X 3

8a. 45·9 X 2   b. 2·44 X 8   c. 7·24 X 5

9. One brick weighs 1·34 kg. How much will 6 bricks weigh?

10. If you bought 3 records at a cost of £5·15 each, how much would you pay altogether?

11. An athlete runs one lap of a running track in an average time of 62·4 seconds. How long would it take him to run 4 laps if he can keep this pace up?

12. A chimney sweep had nine rods each 1·25 metres long. What length would they be when screwed together?

13. A walker takes 8 strides each 0·92 metres long. How far has he travelled?

31 Division of decimals by numbers less than 10

Examples

1. If three cartons of 'Kwango' orange juice cost £3·90, how much does one cost?
   To get the answer divide £3·90 by 3:

   $$\begin{array}{r} \text{£}1\cdot30 \\ \hline 3\,)\overline{3\cdot90} \end{array}$$

   Answer = £1·30

   Notice that the decimal point in the answer is above the decimal point in the question.

2. 36·4 ÷ 4

   $$\begin{array}{r} 9\cdot1 \\ \hline 4\,)\overline{36\cdot4} \end{array}$$

3. 18·88 ÷ 8

   $$\begin{array}{r} 2\cdot36 \\ \hline 8\,)\overline{18\cdot88} \end{array}$$

4. 27·15 ÷ 3

   $$\begin{array}{r} 9\cdot05 \\ \hline 3\,)\overline{27\cdot15} \end{array}$$

Exercise 1

Do these decimal divisions:

1a. 2)16·8   b. 3)21·9   c. 3)30·15   d. 4)16·48

2a. 4)25·24   b. 4)2·52   c. 4)1·40   d. 5)10·55

3a. 3)13·5   b. 3)1·35   c. 3)0·135   d. 3)135·0

4a. 4)126·4   b. 6)84·6   c. 7)165·2   d. 7)16·52

5a. 6)30·24   b. 5)36·85   c. 8)499·2   d. 9)38·07

6a. 66·57 ÷ 7   b. 308·4 ÷ 6   c. 20·56 ÷ 4   d. 75·66 ÷ 6

7a. 57·50 ÷ 5   b. 68·61 ÷ 3   c. 369·39 ÷ 3   d. 58·88 ÷ 8

8a. 307·8 ÷ 9   b. 190·4 ÷ 8   c. 78·40 ÷ 8   d. 110·79 ÷ 9

9. Divide £124·65 into 9 equal parts.

10. Divide a piece of string 14·34 cm long into 3 equal parts.

11. Divide 2·24 kg into 4 equal parts.

12. A man shared £26·25 equally between his 5 sons. How much did each son receive?

13. An ice-cream salesman divided 16·44 kg of ice cream equally into three bowls. How much did he put into each bowl?

14. A joiner sawed a plank of wood 1·55 metres long, into 5 equal pieces. What was the length of each piece of wood?

15. A racing cyclist cycled seven laps in equal times. If his total time was 150·15 seconds, what was his time for each lap?

32 Multiplication and division of decimals by 10, 100 and 1000

Remember

To multiply a decimal number by 10 or 100 or 1000 move the decimal point to the right. One move for multiplication by 10, two moves for multiplication by 100, three moves for multiplication by 1000, and so on.

To divide by these numbers, move the decimal point to the left by the same amounts.

Examples

1. $2 \cdot 57 \times 10 = \mathbf{25 \cdot 7}$

2. $8 \cdot 6 \times 10 = \mathbf{86 \cdot 0}$

3. $10 \times 2 \cdot 57 = \mathbf{25 \cdot 7}$

4. $5 \cdot 227 \times 100 = \mathbf{522 \cdot 7}$

5. $100 \times 52 \cdot 77 = \mathbf{5227 \cdot 0}$

6. $56 \cdot 3526 \times 1000 = \mathbf{56\,352 \cdot 6}$

7. $56 \cdot 3 \div 10 = \mathbf{5 \cdot 63}$

8. $87 \div 10 = \mathbf{8 \cdot 7}$

9. $456 \cdot 2 \div 100 = \mathbf{4 \cdot 562}$

10. $39 \cdot 73 \div 100 = \mathbf{0 \cdot 3973}$

Exercise 1

Find:

1. $3 \cdot 41 \times 10$	2. $10 \times 6 \cdot 32$	3. $7 \cdot 3 \times 10$
4. $5 \cdot 23 \times 10$	5. $5 \cdot 84 \times 10$	6. $3 \cdot 94 \times 10$
7. $100 \times 3 \cdot 83$	8. $2 \cdot 84 \times 100$	9. $4 \cdot 836 \times 100$
10. $100 \times 7 \cdot 339$	11. $10 \cdot 846 \times 1000$	12. $10 \cdot 004 \times 100$
13. $6 \cdot 34 \times 10$	14. $9 \cdot 004 \times 100$	15. $100 \times 34 \cdot 51$
16. $1000 \times 12 \cdot 99$	17. $34 \cdot 22 \times 100$	18. $2 \cdot 847 \times 10$
19. $100 \times 54 \cdot 837$	20. $35 \cdot 836 \times 1000$	21. $57 \cdot 2 \div 10$
22. $741 \cdot 4 \div 10$	23. $83 \cdot 98 \div 10$	24. $846 \div 100$
25. $64 \cdot 8 \div 100$	26. $947 \cdot 6 \div 100$	27. $23 \cdot 9 \div 100$
28. $45 \cdot 8 \div 10$	29. $83 \cdot 6 \div 100$	30. $356 \cdot 1 \div 1000$
31. $1450 \cdot 03 \div 1000$	32. $49 \cdot 002 \div 100$	33. $73 \cdot 725 \div 100$
34. $67 \cdot 946 \div 10$	35. $3 \cdot 746 \div 10$	36. $49 \cdot 007 \div 100$
37. $20 \cdot 737 \div 10$	38. $60 \cdot 047 \div 100$	39. $80 \cdot 0005 \div 10$
40. $2359 \cdot 2 \div 1000$		

33 Decimals revision crossword

This crossword is based on work done on decimals. Copy the crossword and fill in the answers to the clues.

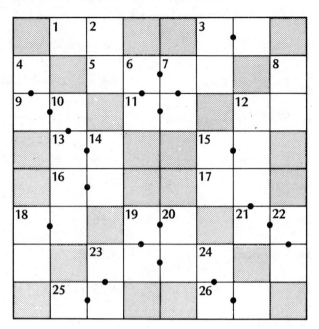

ACROSS

1. Write 16·8 to 2 s.f.
3. 0·6 × 3
5. Write 48·868 to 2 decimal places
9. 4·8 + 3·4
11. (3·8 + 6·4) − 5·7
12. 3·6 × 10
13. Write 3½ as a decimal number
15. 0·8 × 8
16. 3·2 ÷ 2
17. Write 31·366 to 2 s.f.
18. 0·9 × 3
19. Calculate (3·78 + 1·83) to 1 decimal place
21. Write $6\frac{3}{10}$ as a decimal number
23. Write 26·456 to 2 decimal places
25. Find (4·36 + 1·47 + 3·32) to 2 s.f.
26. 26·4 ÷ 8

DOWN

2. 41·5 + 32·5
3. 0·17 × 100
4. 8·3 − 3·5
6. 9·64 − 1·27 to 1 decimal place
7. 3·2 + 5·3
8. 46·362 to 2 s.f.
10. 2·3166 to 4 s.f.
12. 3·416 × 100
14. 0·56 × 100
15. 26·8 + 36·2
18. Calculate (4·2 × 6) to 2 s.f.
19. Write $5\frac{3}{5}$ as a decimal number
20. 6·4 × 10
22. 39 ÷ 10
23. 19·8 ÷ 9
24. 37·8 ÷ 6

34 Multiplication of fractions by whole numbers

Examples

1. What is **half** of £4? The answer is **£2.**
   To get this answer you can **multiply** $\frac{1}{2}$ by £4.
   $\frac{1}{2} \times £4 = \frac{1 \times £4}{2} = \frac{£4}{2} = £2.$

2. Find $\frac{1}{4}$ of £8. $\frac{1}{4} \times £8 = \frac{1 \times £8}{4} = \frac{£8}{4} = £2$

3. Find $\frac{3}{4}$ of £16. $\frac{3}{4} \times £16 = \frac{3 \times £16}{4} = \frac{£48}{4} = £12$

4. Find $\frac{2}{3}$ of 10 kg of flour $\frac{2}{3} \times 10 \text{ kg} = \frac{2 \times 10}{3} \text{ kg} = \frac{20}{3} \text{ kg} = 6\frac{2}{3} \text{ kg}$

Exercise 1

Find the given fraction of each quantity:

1a. $\frac{1}{2}$ of £6 | b. $\frac{1}{2}$ of 8 kg | c. $\frac{1}{2}$ of 110 cm

2a. $\frac{1}{4}$ of 12 kg | b. $\frac{1}{4}$ of 16 l | c. $\frac{1}{4}$ of 220 mm

3a. $\frac{1}{3}$ of 9 cm | b. $\frac{1}{3}$ of 15 ml | c. $\frac{1}{3}$ of 18 g

4a. $\frac{1}{5}$ of 15 m | b. $\frac{1}{5}$ of 25 g | c. $\frac{1}{5}$ of 35 l

5a. $\frac{2}{3}$ of 12 kg | b. $\frac{2}{3}$ of 15 mm | c. $\frac{2}{3}$ of 30 ml

6a. $\frac{3}{4}$ of 16 l | b. $\frac{3}{4}$ of 24 t | c. $\frac{3}{4}$ of 40 km

7a. $\frac{2}{5}$ of 18 ml | b. $\frac{2}{5}$ of 40 m | c. $\frac{2}{5}$ of 58 g

8a. $\frac{3}{5}$ of 20 mm | b. $\frac{3}{5}$ of 35 kg | c. $\frac{3}{5}$ of 52 l

9a. $\frac{3}{10}$ of 40 cm | b. $\frac{3}{10}$ of 60 l | c. $\frac{3}{10}$ of 83 t

10a. $\frac{7}{10}$ of 30 mm | b. $\frac{7}{10}$ of 70 kg | c. $\frac{7}{10}$ of 900 g

11a. $\frac{1}{6}$ of 30 mm | b. $\frac{1}{6}$ of 42 mm | c. $\frac{1}{6}$ of 54 mm

12a. $\frac{5}{6}$ of 18 l | b. $\frac{5}{6}$ of 36 l | c. $\frac{5}{6}$ of 48 l

13a. $\frac{1}{8}$ of 16 mg | b. $\frac{1}{8}$ of 24 mg | c. $\frac{1}{8}$ of 40 mg

14a. $\frac{5}{8}$ of 24 t | b. $\frac{5}{8}$ of 32 t | c. $\frac{5}{8}$ of 48 t

15a. $\frac{7}{8}$ of 16 kg | b. $\frac{7}{8}$ of 56 kg | c. $\frac{7}{8}$ of 80 kg

16a. $\frac{1}{7}$ of 42 t | b. $\frac{3}{7}$ of 84 g | c. $\frac{5}{7}$ of 49 mm

Examples

1. If you buy four, half-kilogram blocks of cheese, how many kilograms would you have bought altogether?
    You should have an answer of 2 kg of cheese. To get this answer you can **multiply** 4 by $\frac{1}{2}$:

    $$4 \times \tfrac{1}{2} \text{ kg} = \tfrac{4 \times 1}{2} \text{ kg} = \tfrac{4}{2} \text{ kg} = \textbf{2 kg}$$

2. Find $8 \times \frac{1}{4}$ cm. $8 \times \tfrac{1}{4} \text{ cm} = \tfrac{8 \times 1}{4} \text{ cm} = \tfrac{8}{4} \text{ cm} = \textbf{2 cm}$

3. Find $5 \times \frac{1}{2}$ tonne. $5 \times \tfrac{1}{2} \text{ tonne} = \tfrac{5 \times 1}{2} = \tfrac{5}{2} = \textbf{2}\tfrac{1}{2} \textbf{ tonnes}$

Exercise 2

Find:

1a. $12 \times \frac{1}{2}$ t	b. $16 \times \frac{1}{2}$ km	c. $27 \times \frac{1}{2}$ l
2a. $8 \times \frac{1}{4}$ m	b. $20 \times \frac{1}{4}$ g	c. $25 \times \frac{1}{4}$ ml
3a. $12 \times \frac{1}{3}$ l	b. $18 \times \frac{1}{3}$ cm	c. $32 \times \frac{1}{3}$ kg
4a. $15 \times \frac{1}{5}$ mm	b. $25 \times \frac{1}{5}$ t	c. $41 \times \frac{1}{5}$ ml
5a. $18 \times \frac{2}{3}$ ml	b. $24 \times \frac{2}{3}$ cm	c. $30 \times \frac{2}{3}$ g
6a. $16 \times \frac{3}{4}$ ml	b. $32 \times \frac{3}{4}$ cm	c. $42 \times \frac{3}{4}$ g
7a. $15 \times \frac{2}{5}$ l	b. $25 \times \frac{2}{5}$ kg	c. $28 \times \frac{2}{5}$ mm
8a. $20 \times \frac{3}{5}$ t	b. $35 \times \frac{3}{5}$ cm	c. $28 \times \frac{3}{5}$ l
9a. $20 \times \frac{3}{10}$ mm	b. $40 \times \frac{3}{10}$ kg	c. $60 \times \frac{3}{10}$ ml
10a. $24 \times \frac{9}{10}$ l	b. $36 \times \frac{9}{10}$ m	c. $41 \times \frac{9}{10}$ g
11a. $12 \times \frac{1}{6}$ cm	b. $24 \times \frac{1}{6}$ cm	c. $30 \times \frac{1}{6}$ cm
12a. $18 \times \frac{5}{6}$ g	b. $36 \times \frac{5}{6}$ g	c. $48 \times \frac{5}{6}$ g
13a. $16 \times \frac{1}{8}$ ml	b. $56 \times \frac{1}{8}$ ml	c. $24 \times \frac{1}{8}$ ml
14a. $32 \times \frac{3}{8}$ m	b. $64 \times \frac{3}{8}$ m	c. $48 \times \frac{3}{8}$ m
15a. $72 \times \frac{5}{8}$ t	b. $40 \times \frac{5}{8}$ t	c. $80 \times \frac{5}{8}$ t
16a. $35 \times \frac{1}{7}$ m	b. $63 \times \frac{3}{7}$ kg	c. $77 \times \frac{5}{7}$ cm

35 Addition and subtraction of fractions

Remember It is easy to add (or subtract) fractions which have the same *denominators* (bottom number):

Examples

1. $\frac{3}{4} + \frac{1}{4} = *?$

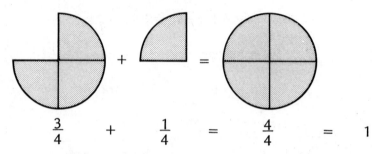

$$\frac{3}{4} \quad + \quad \frac{1}{4} \quad = \quad \frac{4}{4} \quad = \quad 1$$

2. Find $\frac{1}{3} + \frac{1}{3}$ $\frac{1}{3} + \frac{1}{3} = \frac{2}{3}$.

3. Find $\frac{3}{8} + \frac{1}{8}$ $\frac{3}{8} + \frac{1}{8} = \frac{4}{8} = \frac{1}{2}$

4. Find $\frac{2}{5} + \frac{4}{5}$ $\frac{2}{5} + \frac{4}{5} = \frac{6}{5} = 1\frac{1}{5}$

5. Find $\frac{3}{4} + \frac{3}{4}$ $\frac{3}{4} + \frac{3}{4} = \frac{6}{4} = \frac{3}{2} = 1\frac{1}{2}$

Exercise 1 Add together each pair of fractions:

1a. $\frac{1}{2} + \frac{1}{2}$ b. $\frac{1}{4} + \frac{1}{4}$ c. $\frac{2}{5} + \frac{1}{5}$ d. $\frac{3}{8} + \frac{3}{8}$

2a. $\frac{1}{4} + \frac{2}{4}$ b. $\frac{3}{5} + \frac{2}{5}$ c. $\frac{2}{10} + \frac{7}{10}$ d. $\frac{2}{5} + \frac{2}{5}$

3a. $\frac{3}{8} + \frac{5}{8}$ b. $\frac{3}{10} + \frac{1}{10}$ c. $\frac{2}{9} + \frac{4}{9}$ d. $\frac{2}{7} + \frac{3}{7}$

4a. $\frac{2}{6} + \frac{3}{6}$ b. $\frac{2}{5} + \frac{3}{5}$ c. $\frac{2}{3} + \frac{2}{3}$ d. $\frac{5}{8} + \frac{4}{8}$

5a. $\frac{5}{6} + \frac{2}{6}$ b. $\frac{8}{10} + \frac{3}{10}$ c. $\frac{7}{8} + \frac{3}{8}$ d. $\frac{3}{100} + \frac{99}{100}$

Example Find $\frac{5}{8} - \frac{1}{8}$ $\frac{5}{8} - \frac{1}{8} = \frac{4}{8} = \frac{1}{2}$

Exercise 2 Subtract each pair of fractions as shown in the example:

1a. $\frac{3}{4} - \frac{1}{4}$ b. $\frac{4}{5} - \frac{1}{5}$ c. $\frac{3}{5} - \frac{2}{5}$ d. $\frac{4}{5} - \frac{3}{5}$

2a. $\frac{5}{6} - \frac{1}{6}$ b. $\frac{3}{8} - \frac{1}{8}$ c. $\frac{5}{8} - \frac{3}{8}$ d. $\frac{7}{8} - \frac{3}{8}$

3a. $\frac{7}{8} - \frac{5}{8}$ b. $\frac{7}{8} - \frac{1}{8}$ c. $\frac{3}{10} - \frac{1}{10}$ d. $\frac{7}{10} - \frac{3}{10}$

4a. $\frac{7}{10} - \frac{1}{10}$ b. $\frac{9}{10} - \frac{3}{10}$ c. $\frac{9}{10} - \frac{1}{10}$ d. $\frac{66}{100} - \frac{4}{100}$

5a. $\frac{36}{100} - \frac{8}{100}$ b. $\frac{65}{100} - \frac{24}{100}$ c. $\frac{56}{100} - \frac{27}{100}$ d. $\frac{80}{100} - \frac{38}{100}$

36 More addition and subtraction of fractions

Examples

1. $\frac{1}{2} + \frac{1}{4} = *?$

$$\frac{1}{2} \quad + \quad \frac{1}{4} \quad = \quad \frac{2}{4} \quad + \quad \frac{1}{4} \quad = \quad \frac{3}{4}$$

We need to change $\frac{1}{2}$ and write it as $\frac{2}{4}$

Then the denominators of the fractions are the same.

2. Find $\frac{1}{3} + \frac{1}{6}$ $\frac{1}{3} + \frac{1}{6} = \frac{2}{6} + \frac{1}{6} = \frac{3}{6} = \frac{1}{2}$

3. Find $\frac{1}{2} + \frac{3}{4}$ $\frac{1}{2} + \frac{3}{4} = \frac{2}{4} + \frac{3}{4} = \frac{5}{4} = 1\frac{1}{4}$

4. Find $\frac{3}{5} + \frac{1}{10}$ $\frac{3}{5} + \frac{1}{10} = \frac{6}{10} + \frac{1}{10} = \frac{7}{10}$

Exercise 1   Find:

1a. $\frac{1}{4} + \frac{1}{2}$ b. $\frac{3}{4} + \frac{1}{2}$ c. $\frac{2}{3} + \frac{1}{6}$ d. $\frac{3}{10} + \frac{1}{5}$

2a. $\frac{1}{2} + \frac{1}{8}$ b. $\frac{3}{8} + \frac{1}{2}$ c. $\frac{1}{6} + \frac{1}{2}$ d. $\frac{1}{4} + \frac{3}{8}$

3a. $\frac{1}{5} + \frac{7}{10}$ b. $\frac{3}{4} + \frac{1}{8}$ c. $\frac{5}{8} + \frac{1}{4}$ d. $\frac{1}{2} + \frac{3}{10}$

4a. $\frac{7}{10} + \frac{2}{5}$ b. $\frac{3}{8} + \frac{3}{4}$ c. $\frac{2}{3} + \frac{5}{6}$ d. $\frac{1}{2} + \frac{7}{8}$

5a. $\frac{1}{2} + \frac{9}{10}$ b. $\frac{7}{16} + \frac{5}{8}$ c. $\frac{9}{10} + \frac{9}{100}$ d. $\frac{99}{1000} + \frac{1}{10}$

Examples

1. Find $\frac{3}{5} - \frac{1}{10}$ $\frac{3}{5} - \frac{1}{10} = \frac{6}{10} - \frac{1}{10} = \frac{5}{10} = \frac{1}{2}$

2. Find $\frac{3}{4} - \frac{1}{8}$ $\frac{3}{4} - \frac{1}{8} = \frac{6}{8} - \frac{1}{8} = \frac{5}{8}$

Exercise 2   Find:

1a. $\frac{3}{4} - \frac{1}{2}$ b. $\frac{1}{2} - \frac{1}{4}$ c. $\frac{5}{6} - \frac{1}{3}$ d. $\frac{3}{5} - \frac{1}{10}$

2a. $\frac{3}{8} - \frac{1}{4}$ b. $\frac{3}{4} - \frac{5}{8}$ c. $\frac{3}{10} - \frac{1}{5}$ d. $\frac{1}{2} - \frac{1}{6}$

3a. $\frac{1}{4} - \frac{1}{8}$ b. $\frac{1}{2} - \frac{3}{8}$ c. $\frac{2}{5} - \frac{3}{10}$ d. $\frac{7}{10} - \frac{1}{2}$

4a. $\frac{7}{8} - \frac{1}{4}$ b. $\frac{7}{8} - \frac{3}{4}$ c. $\frac{9}{10} - \frac{1}{2}$ d. $\frac{3}{4} - \frac{3}{8}$

5a. $\frac{5}{8} - \frac{1}{2}$ b. $\frac{5}{8} - \frac{1}{4}$ c. $\frac{1}{2} - \frac{1}{8}$ d. $\frac{3}{4} - \frac{1}{8}$

43

37 Addition and subtraction of mixed numbers

Remember

When adding or subtracting mixed numbers you must take the whole number and fractional parts separately:

Examples

1. $2\frac{1}{5} + 1\frac{3}{5} = (2 + 1) + (\frac{1}{5} + \frac{3}{5}) = 3\frac{4}{5}$

2. $5\frac{1}{4} + 2\frac{3}{4} = (5 + 2) + (\frac{1}{4} + \frac{3}{4}) = 7 + 1 = 8$

3. $6\frac{1}{10} + 3\frac{1}{5} = (6 + 3) + (\frac{1}{10} + \frac{1}{5}) = 9 + \frac{3}{10} = 9\frac{3}{10}$

4. $2\frac{2}{5} + \frac{9}{10} = 2 + (\frac{2}{5} + \frac{9}{10}) = 2 + 1\frac{3}{10} = 3\frac{3}{10}$

Exercise 1

Find:

1a. $4\frac{1}{5} + 2\frac{2}{5}$ b. $3\frac{1}{10} + 4\frac{2}{10}$ c. $3\frac{3}{5} + 2\frac{1}{5}$ d. $5\frac{1}{3} + 2\frac{1}{3}$

2a. $2\frac{3}{4} + 6\frac{1}{4}$ b. $5\frac{1}{4} + 5\frac{1}{4}$ c. $6\frac{3}{4} + 2\frac{1}{4}$ d. $10\frac{1}{4} + 5\frac{3}{4}$

3a. $2\frac{1}{10} + 8\frac{5}{10}$ b. $1\frac{1}{8} + 1\frac{4}{8}$ c. $2\frac{1}{2} + 1\frac{1}{4}$ d. $1\frac{1}{2} + 2\frac{1}{4}$

4a. $5\frac{1}{4} + 1\frac{1}{2}$ b. $2\frac{1}{2} + 1\frac{1}{2}$ c. $3\frac{1}{2} + 6\frac{1}{2}$ d. $7\frac{1}{10} + 1\frac{1}{5}$

5a. $2\frac{1}{10} + 2\frac{3}{5}$ b. $1\frac{2}{5} + \frac{7}{10}$ c. $\frac{7}{10} + 2\frac{1}{5}$ d. $1\frac{1}{2} + \frac{3}{4}$

6a. $1\frac{1}{6} + 3\frac{1}{3}$ b. $5\frac{2}{3} + 1\frac{5}{6}$ c. $1\frac{2}{3} + 3\frac{1}{6}$ d. $\frac{5}{6} + 3\frac{1}{3}$

7a. $2\frac{1}{4} + 1\frac{1}{8}$ b. $6\frac{3}{8} + \frac{3}{4}$ c. $1\frac{3}{4} + 4\frac{5}{8}$ d. $2\frac{7}{8} + 4\frac{3}{4}$

8a. $3\frac{1}{6} + 4\frac{1}{12}$ b. $\frac{7}{12} + 2\frac{5}{6}$ c. $2\frac{1}{12} + \frac{5}{6}$ d. $2\frac{1}{6} + 1\frac{7}{12}$

Examples

1. $4\frac{2}{5} - 2\frac{1}{5} = (4 - 2) + (\frac{2}{5} - \frac{1}{5}) = 2\frac{1}{5}$

2. $4\frac{1}{2} - 3\frac{1}{4} = (4 - 3) + (\frac{1}{2} - \frac{1}{4}) = 1 + \frac{1}{4} = 1\frac{1}{4}$

3. $6\frac{2}{5} - 6\frac{1}{10} = (6 - 6) + (\frac{2}{5} - \frac{1}{10}) = 0 + \frac{3}{10} = \frac{3}{10}$

Exercise 2

Find:

1a. $4\frac{2}{3} - 1\frac{1}{3}$ b. $3\frac{3}{4} - 1\frac{1}{4}$ c. $4\frac{3}{5} - 1\frac{2}{5}$ d. $5\frac{4}{5} - 2\frac{1}{5}$

2a. $5\frac{7}{8} - 2\frac{3}{8}$ b. $6\frac{5}{8} - 3\frac{1}{8}$ c. $8\frac{5}{8} - 4\frac{3}{8}$ d. $6\frac{7}{10} - 2\frac{3}{10}$

3a. $3\frac{1}{2} - 1\frac{1}{4}$ b. $6\frac{1}{2} - 2\frac{3}{8}$ c. $5\frac{3}{4} - 1\frac{5}{8}$ d. $8\frac{3}{5} - 3\frac{3}{10}$

4a. $6\frac{1}{2} - 3\frac{3}{10}$ b. $7\frac{3}{5} - 2\frac{1}{10}$ c. $8\frac{4}{5} - 3\frac{1}{10}$ d. $9\frac{3}{4} - 5\frac{3}{8}$

5a. $8\frac{1}{4} - 6\frac{1}{8}$ b. $5\frac{7}{8} - 3\frac{1}{2}$ c. $7\frac{5}{8} - 5\frac{1}{4}$ d. $6\frac{9}{10} - 2\frac{2}{5}$

38 Percentages

Remember

A 'percentage' is a form of fraction. It is a fraction which has a denominator of 100. Instead of writing a fraction you can use the symbol %. 10% is read 'ten per cent'.

Examples

1. $\frac{1}{10} = \frac{10}{100} = $ **10%**
   ten per cent

2. 50% $= \frac{50}{100} = \frac{1}{2}$
   fifty per cent

3. $19\% = \frac{19}{100}$

4. $250\% = \frac{250}{100} = 2\frac{50}{100} = 2\frac{1}{2}$

Exercise 1

Write each of the following percentages as a fraction but then cancel the fraction to its lowest terms, if possible:

1a. 10%	b. 50%	c. 75%	d. 25%
2a. 30%	b. 40%	c. 60%	d. 24%
3a. 5%	b. 90%	c. 80%	d. 45%
4a. 99%	b. 37%	c. 27%	d. 89%
5a. 100%	b. 200%	c. 150%	d. 1000%

6. Write *fifty-five per cent* as a fraction and cancel it to its lowest terms.

Remember

You can also write any fraction as a percentage. If the denominator is 100 then there is no problem. If not, then you must find the equivalent fraction that has the denominator 100.

Examples

1. $\frac{30}{100} = $ **30%**

2. $\frac{55}{100} = $ **55%**

3. $\frac{72}{100} = $ **72%**

4. $\frac{1}{50} = \frac{2}{100} = $ **2%**

5. $\frac{1}{5} = \frac{20}{100} = $ **20%**

6. $\frac{4}{5} = \frac{80}{100} = $ **80%**

Exercise 2

Write each fraction as a percentage:

1a. $\frac{9}{100}$	b. $\frac{19}{100}$	c. $\frac{37}{100}$	d. $\frac{99}{100}$
2a. $\frac{3}{10}$	b. $\frac{5}{10}$	c. $\frac{7}{10}$	d. $\frac{9}{10}$
3a. $\frac{2}{5}$	b. $\frac{3}{5}$	c. $\frac{4}{5}$	d. $\frac{5}{5}$
4a. $\frac{1}{4}$	b. $\frac{3}{4}$	c. $\frac{3}{20}$	d. $\frac{8}{20}$
5a, $\frac{13}{20}$	b. $\frac{17}{20}$	c. $\frac{3}{25}$	d. $\frac{7}{25}$
6a. $\frac{14}{25}$	b. $\frac{21}{25}$	c. $\frac{3}{50}$	d. $\frac{11}{50}$

39 Using percentages

Remember

You will frequently need to find a percentage of a sum of money, or some other quantity:

Examples

1. Find 10% of £30

$$10\% \text{ of } £30 = \frac{10}{100} \times £30$$
$$= \frac{1}{10} \times £30$$
$$= \frac{£30}{10} = £3$$

2. Find 20% of 650 km

$$20\% \text{ of } 650 \text{ km} = \frac{20}{100} \times 650 \text{ km}$$
$$= \frac{1}{5} \times 650 \text{ km}$$
$$= \frac{650}{5} \text{ km} = \textbf{130 km}$$

Exercise 1

Calculate each percentage of the quantity given:

1a. 10% of £50	b. 10% of £70	c. 10% of £10
2a. 50% of £6	b. 50% of £10	c. 50% of 10 km
3a. 25% of £8	b. 25% of £80	c. 25% of 40 women
4a. 75% of £20	b. 75% of £100	c. 75% of 40 men
5a. 30% of £90	b. 70% of £90	c. 100% of 1 workforce

6. If your income is £80 a week then the income tax you will have to pay will be:
   30% of £80
   a. How much income tax do you pay a week?
   b. How much money does that leave you each week?

7. Michael Mole needed to put a 20% deposit on a £15 000 house. How much did he need to put down?

8. A school of 1000 pupils was facing a 'flu epidemic. 25% of the pupils were suffering from 'flu. How many was this?

9. If your income is £80 a week and you are offered a wage rise of 10%, how much extra would you expect to get each week?

10. Out of a class of 28 pupils, 75% passed their end-of-term maths exam. How many passed the exam?

11. A man bought a house for £20 000 and later sold it for a profit of 20%. What was the selling price?

12. A boy bought his motor cycle for £800 and two years later sold it making a loss of 20%. How much did he sell it for?

This crossword is based on work done on fractions. Copy the crossword and fill in the answers to the clues.

1 5		2 1	3 7	0		4 1
5 2	6 5		2		7 1	8
	8 1	9 1		10 2	0	
11 3		12 2	0	0		13 1
	14 1	5		15 3	16 2	
17 2	1		18 5		19 3	20 6
0		21 1	0	6		0

ACROSS

2. Find a quarter of 680.

5. Write $\frac{1}{4}$ as a percentage.

7. Find 50% of 36.

8. The top line of $5\frac{1}{2}$ as an improper fraction.

10. $\frac{5}{4} \times 16$.

11. $\frac{6}{4} = \frac{*}{2}$.

12. Find $\frac{1}{3}$ of 600.

13. 20% of 20.

14. Write $\frac{3}{20}$ as a percentage.

15. Find two-thirds of 48.

17. $\frac{3}{4}$ of 28.

19. $\frac{1}{6} = \frac{6}{**}$.

21. 50% of 212.

DOWN

1. $\frac{52}{100}$ as a percentage.

3. A half of 144.

4. $\frac{9}{2} = \frac{**}{4}$

6. $6\frac{3}{8}$ as an improper fraction is $\frac{**}{8}$.

7. $1\frac{3}{7} = \frac{**}{7}$.

9. $1\frac{25}{100} = \frac{***}{100}$

10. One third of 609.

14. $\frac{33}{9} = \frac{**}{3}$.

16. Half of 46.

17. 10% of 200.

18. Write $\frac{1}{2}$ as a percentage.

20. 60% of 100.

1. Write down these numbers in words:

   a. 19 Ninteen   b. 42 forty two   c. 265 two h Sixty d. 34013 thu* 1   e. 13 409 13 thu* 9

2. Find the value of * in the following:

   a. $* + 6 = 11$ 5   b. $* - 8 = 4$

   c. $11 + * = 21$ 10   d. $14 - * = 4$

   e. $* + 8 = 16$ f. $* - 12 = 9$

   g. $16 + * = 32$ h. $21 - * = 15$

3a. A man changes £75 into ten pence pieces to use in fruit machines. How many coins will he receive?

   b. After playing for a few hours, the man has 360 ten pence pieces left. He has this changed into pound notes. How many will he receive?

4a.   16   b.   36   c.   104   d.   166   e.   341

   $\times\ 6$ $\times\ 7$ $\times\ 8$ $\times\ 9$ $\times\ 5$

5. Copy and complete the following fractions:

   a. $\frac{3}{5} = \frac{9}{**}$ b. $\frac{7}{8} = \frac{**}{32}$ c. $\frac{16}{20} = \frac{*}{5}$ d. $\frac{4}{9} = \frac{16}{**}$ e. $\frac{24}{36} = \frac{2}{*}$

6. Cancel down each of the following fractions to its lowest terms:

   a. $\frac{18}{24}$ b. $\frac{20}{40}$ c. $\frac{24}{42}$ d. $\frac{27}{63}$ e. $\frac{36}{72}$

7. Change each improper fraction into a mixed number:

   a. $\frac{10}{3}$ b. $\frac{19}{4}$ c. $\frac{27}{5}$ d. $\frac{31}{5}$ e. $\frac{40}{6}$

8. Arrange each set of decimal numbers in order of size, smallest first:

   a. 2·6, 8·2, 5·4, 4·6, 4·5

   b. 6·801, 6·9, 6·866, 6·7

   c. 3·421, 3·42, 3·424, 3·43

9. Write the following decimals as fractions then cancel them down where possible:

   a. 0·4   b. 0·13   c. 0·21   d. 0·26   e. 0·32

   f. 0·50   g. 0·112   h. 0·128   i. 0·33   j. 0·625

10. Change the following fractions into decimals:

   a. $\frac{1}{5}$ b. $\frac{1}{7}$ c. $\frac{2}{5}$ d. $\frac{3}{8}$ e. $\frac{7}{10}$

11. Change the following mixed numbers into decimals:

   a. $2\frac{3}{5}$ b. $3\frac{3}{10}$ c. $5\frac{3}{25}$ d. $6\frac{13}{25}$ e. $8\frac{19}{50}$

12. Write each decimal number to the given number of places:

   a. 3·28   to 1 d.p.   b. 4·66   to 1 d.p.

   c. 1·42   to 1 d.p.   d. 6·314   to 2 d.p.

   e. 2·127   to 2 d.p.   f. 1·604   to 2 d.p.

   g. 3·832   to 1 d.p.   h. 1·666   to 1 d.p.

13. Write each decimal number to the given number of significant figures:

    a. 1·46 to 2 s.f.   b. 1·31 to 2 s.f.
    c. 6·412 to 3 s.f.   d. 7·862 to 3 s.f.
    e. 10·66 to 3 s.f.   f. 12·41 to 2 s.f.
    g. 12·40 to 2 s.f.   h. 3·004 to 1 s.f.
    i. 11·02 to 2 s.f.   j. 9·8090 to 3 s.f.

14a.
$$6·8 + 4·6$$
b. $5·9 + 6·8$ c. $3·62 + 1·66$ d. $4·85 + 2·68$ e. $7·662 + 1·288$

15a.
$$9·6 - 5·3$$
b. $6·4 - 3·7$ c. $6·24 - 3·82$ d. $15·774 - 13·19$ e. $29·23 - 26·756$

16a. Divide £110·40 equally among six people.
    b. A runner ran seven laps in 464·8 seconds. If he ran each lap in the same time, what was this time?
    c. Divide £147·78 by nine.

17. Work out the following:

    a. 6·42 × 10   b. 13·6 × 10
    c. 1·41 × 100   d. 6·664 × 100
    e. 1·7428 × 1000   f. 4·864 ÷ 10
    g. 8·66 ÷ 10   h. 384·2 ÷ 100
    i. 6667 ÷ 100   j. 42 866 ÷ 1000

18. Find the value of the following:

    a. $\frac{1}{2}$ of £14   b. $\frac{1}{3}$ of £15
    c. $\frac{1}{5}$ of 35 kg   d. $\frac{2}{5}$ of 65 m
    e. $\frac{3}{8}$ of 128 mm   f. $\frac{5}{8}$ of 640 l
    g. $\frac{3}{10}$ of £80   h. $\frac{7}{10}$ of 40 g
    i. $\frac{1}{20}$ of 300 cm   j. $\frac{1}{50}$ of 1000 m

19a. $\frac{1}{5} + \frac{2}{5}$ b. $\frac{3}{8} + \frac{3}{8}$ c. $\frac{3}{10} + \frac{3}{10}$ d. $\frac{7}{8} - \frac{3}{8}$ e. $\frac{4}{5} - \frac{2}{5}$

20a. $3\frac{1}{4} + 4\frac{3}{8}$ b. $2\frac{3}{5} + 1\frac{7}{10}$ c. $3\frac{9}{10} + 4\frac{2}{5}$ d. $5\frac{5}{8} - 4\frac{1}{4}$ e. $9\frac{4}{5} - 6\frac{3}{10}$

21. Calculate each percentage:

    a. 10% of £60   b. 50% of £17   c. 25% of £18   d. 75% of £24   e. 60% of £40
    f. 40% of £15   g. 15% of £40   h. 44% of £300   i. 5% of £80   j. 100% of £10

22. If your weekly wage was £65 per week and you were offered a rise of 20%, what would be your new wage?

23. A man bought a house for £20 000 and promised to pay 15% deposit. How much was this deposit?

24. 120 pupils in the fifth year took a Maths exam and 20% failed. Find how many passed.

41 Money

Examples

1. £1 and 32 pence = **£1·32**
2. £12 and 3 pence = **£12·03**
3. £8 and 30 pence = **£8·30**

Exercise 1   Rewrite each sum of money as a decimal number:

1a. £2 and 64 pence   b. £4 and 14 pence

2a. £16 and 13 pence   b. £42 and 20 pence

3a. £800 and 46 pence   b. £6 and 99 pence

4a. £38 and 20 pence   b. £10 and 35 pence

5a. £400 and 65 pence   b. £326 and 88 pence

6a. £5 and 4 pence   b. £8 and 9 pence

7a. £22 and 1 penny   b. £0 and 2 pence

8a. £16 and 8 pence   b. £0 and 16 pence

Examples   1. £2·35 = **235p**   2. £0·05 = **5p**

Exercise 2   Express each sum of money in pence:

1a. £0·00	b. £0·01	c. £0·03	d. £0·09
2a. £0·16	b. £0·35	c. £0·48	d. £0·66
3a. £0·75	b. £0·86	c. £0·90	d. £0·96
4a. £1·09	b. £1·62	c. £2·02	d. £3·65
5a. £4·69	b. £5·78	c. £6·82	d. £9·16
6a. £4·66	b. £5·80	c. £1·66	d. £4·77
7a. £5·84	b. £2·94	c. £1·91	d. £0·36
8a. £3·06	b. £3·82	c. £3·62	d. £0·14

Examples   1. 1363p = **£13·63**   2. 8p = **£0·08**

3. 50p = **£0·50**   4. 105p = **£1·05**

Exercise 3   Express each sum of money in pounds:

1a. 7p	b. 15p	c. 48p	d. 96p
2a. 125p	b. 264p	c. 381p	d. 430p
3a. 506p	b. 645p	c. 726p	d. 836p
4a. 945p	b. 1012p	c. 1141p	d. 1270p

5a. 1365p	b. 1414p	c. 1563p	d. 1658p
6a. 1832p	b. 2004p	c. 2104p	d. 2320p
7a. 2307p	b. 2302p	c. 3175p	d. 3240p
8a. 4416p	b. 5004p	c. 6386p	d. 8041p

Examples

1.	one pound forty-six pence	= **£1·46**
2.	three pounds forty pence	= **£3·40**
3.	eight pounds and five pence	= **£8·05**
4.	fourteen pence	= **£0·14**

Exercise 4   Write each sum of money as a decimal number:

1a. two pounds thirty-nine pence	**b.** one pound twenty pence
2a. twelve pounds ninety-one pence	**b.** three pounds twelve pence
3a. seven pounds and two pence	**b.** four pounds and nine pence
4a. one pound and one penny	**b.** five pounds and seven pence
5a. thirty-seven pence	**b.** fifty pence
6a. three pence	**b.** one penny
7a. ninety-nine pence	**b.** ten pounds and six pence
8a. fifteen pounds sixty-three pence	**b.** twenty pounds fifty pence

Examples

1.	£2·05	= **two pounds and five pence**
2.	£5·10	= **five pounds ten pence**
3.	£0·04	= **four pence**

Exercise 5   Write each sum of money in words:

1a. £7·45	b. £5·45	c. £7·30	d. £2·71
2a. £5·39	b. £3·90	c. £3·05	d. £5·09
3a. £0·38	b. £0·60	c. £0·06	d. £0·00
4a. £3·20	b. £8·38	c. £4·06	d. £2·41
5a. £4·11	b. £6·22	c. £3·63	d. £4·84
6a. £2·12	b. £8·04	c. £9·17	d. £6·29
7a. £11·19	b. £13·25	c. £15·58	d. £18·67
8a. £26·69	b. £31·75	c. £42·80	d. £50·99

42 Addition and subtraction of money

Remember Addition (or subtraction) of money is done in exactly the same way as the addition (or subtraction) of decimal numbers.

Examples

1. £1·06
   +£1·91
   ────
   £2·97

2. £4·70
   +£3·68
   ────
   £8·38

3. £5·59
   +£4·62
   ────
   £10·21

4. £1·30 + £0·38 =   £1·30
                      +£0·38
                      ────
                      £1·68

Exercise 1

1a. £1·90
    +£1·07
    ────

b. £5·73
   +£2·21
   ────

c. £13·50
   +£ 5·49
   ────

2a. £2·65
    +£1·28
    ────

b. £8·25
   +£1·66
   ────

c. £24·61
   +£ 5·29
   ────

3a. £2·47
    +£2·71
    ────

b. £4·67
   +£0·92
   ────

c. £0·74
   +£0·35
   ────

4a. £5·47
    +£3·88
    ────

b. £3·75
   +£1·55
   ────

c. £17·77
   +£ 3·57
   ────

5a. £1·45 + £2·34   b. £7·55 + £2·36   c. £14·85 + £3·77

Examples

1. £2·40
   −£1·30
   ────
   £1·10

2. £4·16
   −£1·05
   ────
   £3·11

3. £3·45
   −£1·38
   ────
   £2·07

4. £3·71
   −£1·80
   ────
   £1·91

5. £7·50
   −£5·47
   ────
   £2·03

6. £3·16
   −£1·59
   ────
   £1·57

Exercise 2

1a. £5·38
    −£2·27
    ────

b. £4·75
   −£1·55
   ────

c. £18·53
   −£ 7·31
   ────

2a. £5·43
    −£1·29
    ────

b. £3·63
   −£3·57
   ────

c. £24·28
   −£ 0·42
   ────

3a. £8·41
    −£2·65
    ────

b. £10·38
   −£ 4·29
   ────

c. £21·80
   −£ 2·91
   ────

4a. £6·35 − £4·27   b. £8·65 − £3·46   c. £10·66 − £7·99

5a. £14·44 − £7·77   b. £35·86 − £34·87   c. £19·79 − £13·99

6a. £12·64 − £8·96   b. £26·03 − £16·58   c. £21·62 − 19·83

43 Multiplication and division of money

Examples

1. £2·31
   X   3
   ——————
   £6·93

2. £2·15
   X   4
   ——————
   £8·60

3. £4·67
   X   6
   ——————
   £28·02

Exercise 1

Work out the following:

1a. £3·41
    X   2

b. £3·23
   X   3

c. £2·70
   X   3

d. £1·25
   X   5

2a. £4·61
    X   3

b. £3·48
   X   2

c. £5·56
   X   2

d. £0·73
   X   7

3a. £6·06
    X   6

b. £5·89
   X   4

c. £3·66
   X   8

d. £4·08
   X   9

4. 'Bertrand's' Boot Polish costs £0·36 a tin. How much would eight tins cost?

5. A good piece of steak costs £4·53 for 1 kilogram. How much would 3 kg cost?

Examples

1. £12·21 ÷ 3

   £ 4·07
   ————————
   3) 12·21   so, £12·21 ÷ 3 = **£4·07**

2. £71·20 ÷ 8

   £ 8·90
   ————————
   8) 71·20   so, £71·20 ÷ 8 = **£8·90**

Exercise 2

Work out the following:

1a. £6·39 ÷ 3   b. £82·26 ÷ 2   c. £4·08 ÷ 4

2a. £3·60 ÷ 6   b. £36·00 ÷ 9   c. £14·91 ÷ 7

3a. £12·60 ÷ 4   b. £15·50 ÷ 5   c. £8·10 ÷ 9

4a. £5·04 ÷ 8   b. £7·14 ÷ 7   c. £25·74 ÷ 9

5. A set of 6 table mats costs £3·72. What is the cost of each table mat?

6. The weight of 5 identical cans of soup is 2·25 kg. How much does one can weigh?

7. Divide £45·04 into 8 equal shares.

8. Seven boys cut a rope of length 15·05 m into equal parts. What length does each boy receive?

9. Fourteen women of equal weight stand on a large set of scales. The total is 2133·6 kg. Find the weight of each woman.

44 Pay and overtime

Remember   Most people work a set number of hours each week — between 35 and 40 hours. If you work more than this you can expect to receive *overtime* rates of pay. For each extra hour of work you are paid *as if you had worked* $1\frac{1}{4}$, $1\frac{1}{2}$ or 2 hours.

Example

normal rate of pay	"time and a quarter" (for extra work during the week)	"time and a half" (for extra work on Saturdays)	"double time" (for extra work on Sunday)
160p per hour	$160 + (\frac{1}{4}$ of $160)$ $= 160 + 40$ $= \mathbf{200p}$	$160 + (\frac{1}{2}$ of $160)$ $= 160 + 80$ $= \mathbf{240p}$	160×2 $= \mathbf{320p}$

Exercise 1   Copy this table into your book and complete it. You will need it later.

rate of pay	time and a quarter	time and a half	double time
80p	100p	120 p	160 p
100p			
120p			
140p			
160p			
180p			
200p			
220p			
240p			
260p			
280p			
300p			
320p			
340p			
360p			
380p			
400p			
420p			
440p			
460p			
480p			
500p			

45 Gross wages

Remember
A *gross wage* is an amount that somebody earns before any deductions are made.

Example
A joiner earns 220p an hour. He works 40 hours in a week.
Find his gross wage.

Gross wage =   220
             X 40
             ─────
             8800p

         = **£88·00**

Exercise 1

1. 250p X 40   2. 360p X 40

3. 125p X 40   4. 178p X 40

5. 265p X 30   6. 375p X 30

7. 340p X 35   8. 450p X 38

9. 244p X 35   10. 356p X 38

Exercise 2

Find the gross wages of each of the following workers:

1. An electrician who works 40 hours at 275p per hour.

2. A builder who works 40 hours at 420p per hour.

3. A plasterer who works 40 hours at 460p per hour.

4. A car mechanic who works 40 hours at 240p per hour.

5. A shop assistant who works 40 hours at 190p per hour.

6. A baker who works 40 hours at 185p per hour.

7. A carpenter who works 40 hours at 214p per hour.

8. A secretary who works 38 hours at 180p per hour.

9. An office boy who works 35 hours at 122p per hour.

10. A bus driver who works 35 hours at 230p per hour.

11. A plumber who works 36 hours at 280p per hour.

12. A painter who works 35 hours at 236p per hour.

13. A bus conductor who works 36 hours at 212p per hour.

14. A butcher who works 34 hours at 266p per hour.

46 Gross wages: overtime

Remember

If you work more than 40 hours in a week you will usually receive overtime pay at *"time and a quarter"*. You should use your overtime rates from the table done earlier.

Example

A plumber works 46 hours in one week at normal rates of 220p per hour. Find his gross wage for the week.

Normal pay =   220   Overtime =   275
                  $\times$ 40 $\times$ 6

            8800p = £88·00   1650p = £16·50

Gross wage = £88·00 + £16·50
            = **£104·50**

Exercise 1

Find the gross wage of the following workers. They all receive "time and a quarter" after 40 hours work:

1. A shop assistant who works 50 hours at 180p per hour.
2. A telephonist who works 48 hours at 140p per hour.
3. A road sweeper who works 46 hours at 220p per hour.
4. A refuse collector who works 50 hours at 260p per hour.
5. A farm worker who works 50 hours at 280p per hour.
6. A motor cycle mechanic who works 52 hours at 280p per hour.
7. An engineer who works 50 hours at 420p per hour.
8. A television engineer who works 44 hours at 440p per hour.
9. An ice cream van driver who works 60 hours at 220p per hour.
10. A journalist who works 56 hours at 480p per hour.

47 Gross wages: pay slip

Example   Here is Joe Brown's pay slip:

normal pay 160p/hr.			overtime paid as 'time and a quarter'. Saturday: 'time and a half'			
day	hours clocked	normal hours	overtime hours	normal pay = rate $\times$ hrs	overtime = over rate $\times$ hrs	Total Pay
MONDAY	10	8	2	1280p	400p	£16·80
TUESDAY	11	8	3	1280p	600p	£18·80
WEDNESDAY	8	8	0	1280p	-	£12·80
THURSDAY	12	8	4	1280p	800p	£20·80
FRIDAY	8	8	0	1280p	-	£12·80
SATURDAY	3	0	3		720p	£ 7·20
SUNDAY						
					gross wage	£89·20

Exercise 1   **1.** Copy and complete Gary Green's pay slip:

normal pay 100p/hr		week day overtime: 'time and a quarter'. Saturday: 'time and a half' Sunday: 'double time				
day	hours clocked	normal hours	overtime hours	normal pay	overtime	total pay
MONDAY	11	8				
TUESDAY	8	8				
WEDNESDAY	10	8				
THURSDAY	12	8				
FRIDAY	8	8				
SATURDAY	4	0	4			
SUNDAY						
					gross wage	

2. Copy and complete Buster Brown's pay slip:

normal pay 240p/hr		week day overtime: 'time and a quarter'. Saturday: 'time and a half' Sunday: 'double time'				
day	hours clocked	normal hours	overtime hours	normal pay	overtime	total pay
MONDAY	10					
TUESDAY	8					
WEDNESDAY	11					
THURSDAY	10					
FRIDAY	10					
SATURDAY						
SUNDAY	6	0	6			
					gross wage	

3. A builder earns 360p per hour. Make up a pay slip like the ones above if he works 10 hours on Monday, 12 hours on Tuesday, 14 hours on Wednesday, 12 hours on Thursday and 10 hours on Saturday.

48 Piecework

Remember

Instead of paying at an hourly rate, some firms pay each worker according to the number of articles that worker has made. This is called *piecework*.

Examples

1.

name	operation	rate	number
Javitt	drilling	8p each	190

Javitt's wage for the day = 8p × 190 = 1520p = **£15·20**

2.

name	operation	rate	number
Johnson	packing	12p for 10	1500

Johnson's wage for the day is 12p × (1500 ÷ 10)
= 12p × 150 = 1800p = **£18·00**

Exercise 1

Work out the following:

1a. 2p × 60   b. 3p × 100   6a. 6p × 388   b. 7p × 462

2a. 4p × 150   b. 5p × 250   7a. (2p for 10) × 50   b. (3p for 10) × 100

3a. 6p × 340   b. 7p × 400   8a. (4p for 10) × 150   b. (5p for 10) × 200

4a. 2p × 75   b. 3p × 144   9a. (5p for 10) × 340   b. (6p for 10) × 360

5a. 4p × 225   b. 5p × 308   10a. (6p for 10) × 440 b. (7p for 10) × 480

Exercise 2

Calculate each person's daily wage:

name	operation	rate	number completed
Armstrong	turning	8p each	144
Clark	painting	6p each	185
Dawson	drilling	4p each	224
Elliot	sewing	5p each	350
Gough	ironing	7p each	244
Harkness	cutting	8p each	228
Leniham	steaming	9p each	195
Miller	grinding	6p each	186
Pratt	hammering	9p each	234
Rose	polishing	6p each	260
Singh	fitting	(20p for 10)	450
White	ironing	(30p for 10)	300
Winter	dyeing	(35p for 10)	400
Yardley	cutting	(40p for 10)	240
Young	drilling	(50p for 10)	320

49 Wages: salaries

Remember   People such as Government officials, teachers and businessmen are paid a certain amount for the year's work. They do not receive overtime and usually get paid at the end of each month. This payment is called a *salary*.

Example   A schoolteacher earns £5400 per year. What is his gross salary each month?

Gross salary each month = £5400 ÷ 12
= **£450**

Exercise 1   Find the monthly salary of the following workers:

1. A private secretary who earns £4800 per year.
2. A lawyer who earns £8040 per year.
3. A staff nurse who earns £7680 per year.
4. A headmaster who earns £10 800 per year.
5. A trainee accountant who earns £4860 each year.
6. A fireman who earns £6240 each year.
7. A typist who earns £3960 per year.
8. A journalist who earns £7920 per year.
9. A clerk who earns £3360 each year.
10. A prison officer who earns £9000 per year.

Example   A doctor receives a monthly salary of £680 per month. How much does he receive each year?

Annual Salary = £680 × 12
= **£8160**

Exercise 2   Find the annual salary of the following people:

1. A doctor with a monthly salary of £960.
2. A dentist with a monthly salary of £920.
3. A teacher with a monthly salary of £554.
4. A prison officer with a monthly salary of £735.
5. A policeman with a monthly salary of £565.
6. A lawyer with a monthly salary of £865.
7. A government inspector with a monthly salary of £955.
8. A car salesman with a monthly salary of £584.
9. A building society manager with a monthly salary of £930.
10. A bank manager with a monthly salary of £922.

50 Net wage

The amount of money that you earn is called your *gross wage*. But before you receive your pay, the following deductions are made.
1. Superannuation (to go towards your pension)
2. National Insurance contribution (for the department of Health and Social Security)
3. Income Tax (for the Government)

Your 'take home pay' is the amount left after these deductions. This is called your 'net wage'.

Example

A joiner's gross wage for one week is £110. He pays £25 income tax, £6 superannuation and £7·50 National Insurance.
Find his net wage.

Gross wage = £110·00   Deductions =   £25·00
                                                   6·00
                                              +   7·50
                                                 ‾‾‾‾‾‾
                                                 £38·50

Net wage =   £110·00
          −   38·50
              ‾‾‾‾‾‾‾
              £71·50

Exercise 1

Find out the net wages of the following people:

	occupation	gross wage	superannuation	N.I.	income tax
1.	fireman	£118	£7·08	£8·00	£18·60
2.	chemist	£146	£8·76	£9·85	£14·40
3.	grinder	£128	£7·68	£8·60	£10·80
4.	turner	£140	£8·40	£9·40	£16·40
5.	electrician	£126	£7·56	£8·50	£22·60
6.	builder	£142	£8·52	£9·60	£15·80
7.	plasterer	£136	£8·16	£9·20	£20·40
8.	farmhand	£98	£5·88	£6·60	£16·40
9.	lorrydriver	£160	£9·60	£10·80	£27·50
10.	car worker	£135	£8·10	£9·10	£40·00
11.	plumber	£138	£8·28	£9·45	£16·40
12.	joiner	£155	£9·30	£10·30	£12·60
13.	painter	£134	£8·04	£9·10	£10·90
14.	steelworker	£170	£10·20	£11·10	£15·50
15.	toolsetter	£175	£10·50	£11·25	£18·70

51 Income tax: P.A.Y.E.

Remember

P.A.Y.E. stands for 'Pay As You Earn'. First, your contributions for Superannuation and National Insurance are taken off. Then you are given some *allowances* which are not taxed. On the remainder of your wage, tax is charged at 30p in the pound.

annual allowances		
married	over 65 under 65	£4135 £3065
single	over 65 under 65	£2600 £1965
housekeeper		£130
dependent relative		£130

Example

A married man aged 45 earns £3500 after Superannuation and N.I. contributions. His mother in law lives with him.

Find a)   the total annual tax allowance
　　 b)   the taxable pay
　　 c)   the annual tax.

a) Allowances =　£3065
　　　　　　　+ £130
　　　　　　　――――――
　　　　　　　£3195

b) Taxable pay =　£3500
　　　　　　　　−£3195
　　　　　　　　――――――
　　　　　　　　£305

c) Annual tax = 305 × 30p
　　　　　　　 = 9150p
　　　　　　　 = **£91·50**

Exercise 1

The earnings used in this exercise are after offtakes for Superannuation and National Insurance. Find a) b) and c) as in the example.

1. Bob Thrift earns £3600. He is single and aged 35.

2. Mary Brown earns £6500. She is single and aged 33.

3. Jimmy Whizz earns £5000. He is married and aged 45.

4. Fred Doolittle earns £8500. He is married, aged 45 and has his old mother living in with him and his wife.

5. Percy Main earns £6500. He is married, aged 55 and has a housekeeper.

6. Jack Green is 67 years old. He is single and owns a small corner shop. His annual earnings are £4800.

7. Sarah Black is single; she is 56 years old. She has her mother living with her and employs a housekeeper. Her earnings are £8765.

Remember

Provided that you have had a job a sufficient length of time, if you go out of work or your income drops, you become entitled to supplementary benefit. The following table shows the benefits allowed per week:

married couple	£49·40	blindness	£1·80
single person (18 or over)	£30·40		
dependent children		extra heating	£4·00
under 11	£10·40		
11–15	£15·60	special diet	£4·00
16–17	£18·70		
18 and over	£24·35	mortgage: interest only	
rent	full amount		

Examples

1. Find the supplementary benefits entitled to a married couple who are out of work and have a dependent daughter aged 13.

   Married couple   £49·40
   Dependent child   £15·60

         Total   **£65·00**

   If a person has a small income then the benefits are reduced by that amount.

2. The same example as above except that the couple have a small income of £15·40.

   Benefits   £65·00
   Income   £15·40

   Net amount   £49·60

Exercise 1

Find the total amount of supplemantary benefit paid out to:

1. A married couple out of work who have two children, one aged 13 and the other aged 15. They pay £21·50 rent.

2. A single person out of work who needs extra heating and pays rent of £15·20.

3. A married couple out of work, one being blind, needing extra heating and paying £8·60 rent.

4. A single person who pays a rent of £10·80 and needs special food.

5. A married couple out of work with four sons aged 6, 8, 12 and 16 years of age.

6. A married couple with three daughters aged 15, 16 and 19, one of them on a special diet. The couple pay £16·50 rent each week and receive an extra heating allowance. Their income is £12·60 per week.

7. A single person who pays £16·40 rent and needs an extra heating allowance. He has an income of £20·80 each week.

8. A married couple who pay £16·10 rent. They have a blind daughter aged 16 and a dependent son aged 17 who is on a special diet. Their total income is £16·00.

9. A married couple with six children aged 7, 9, 9, 10, 12 and 17. They need extra heating and have an income of £22·90.

10. A married couple paying £18·50 interest on their mortgage loan with twins aged 15 and a son aged 20 who is dependent on them. Their income is £14·60.

Remember

When you leave school, if you are unable to get a job, you receive unemployment benefit.

age	living with parents	living alone
16 yr	£17·85	£20·85
17 yr	£17·85	£20·85
18 yr	£22·25	£25·25
		rent: full amount

Example

Gary Green leaves school at 17 and can't get a job. Find his unemployment benefit if he lives with his parents.
Age — 17: Benefit — **£17·85**

Exercise 2

Find the benefits paid to the following people who leave school and can't get a job:

1. Sheila Jackson, aged 16, living with parents.

2. Billy Blacklaw, aged 18, living with parents.

3. Stella Richards, aged 17, living with parents.

4. Jimmy Jones, aged 18, living in a flat with the rent costing £18·60.

5. Norma Bell, aged 18, living in a flat with the rent costing £20·80.

6. How much benefit will Sheila Jackson recieve in a year (52 weeks)?

53 Banking: paying in

Remember

A customer can open a Bank account by depositing any sum at the Bank. When you pay in money it is a good idea to keep a record of how much you have. To help with this you are given a paying in book. A page looks like the one below:

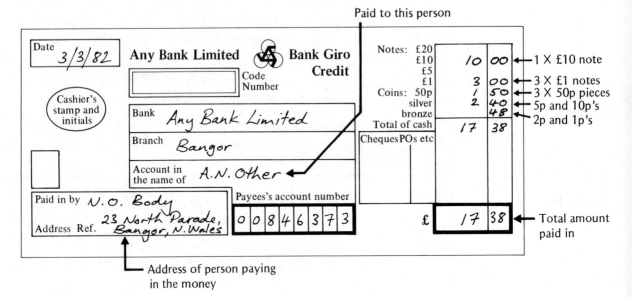

Exercise 1

Draw a copy of the paying in slip into your book, then fill it in using the information given below. Use today's date, pay it to a friend and use your own name and address.

1. 2 X £10 notes, 3 X £5 notes, 12 X £1 notes, 4 X 50p pieces, £3·40 in silver and 70p in copper.

   In the following questions, just copy the right hand side, filling in the 'total amount paid in' box.

2. 2 X £20, 3 X £5, 8 X £1, 2 X 50p, £1·25 in silver and 18p in copper.

3. 1 X £20, 7 X £10, 20 X £1, 7 X 50p, £2·60 in silver, 26p in copper and a cheque for £3·27.

4. 3 X £5, 14 X £1, 10 X 50p, 65p in silver and 12p in copper, one cheque for £24·16 and a Postal Order for £17·50.

5. 3 X £20, 6 X £5, 8 X £1, 13 X 50p, 8 X 10p, 13 X 2p, cheques for £13·86 and £4·68 and a postal order for £8·92.

6. 4 X £20, 2 X £10, 6 X £5, 3 X 50p, 6 X 10p, 12 X 2p, a postal order for £18·62.

7. 6 X £20, 5 X £10, 8 X £5, 7 X 10p, 22 X 5p, two cheques for £3·62 and £42·81.

8. 4 X £10, 22 X £5, 16 X 50p, 12 X 10p, £4·80 in silver, 17 X 2p, a cheque for £18·76 and a postal order for £6·85.

Remember

When you put money into a Bank, you are given an account number and a cheque book. You can use it to take money out of your account and also to pay out money to other people.

A cheque looks like this:

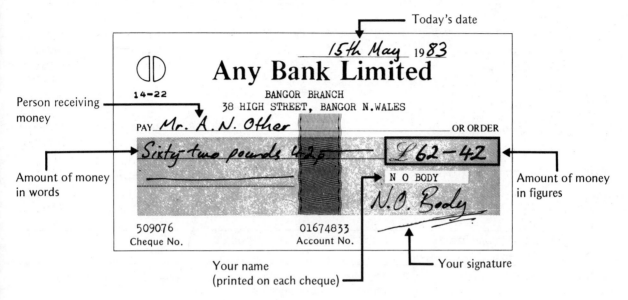

Today's date

Person receiving money

Amount of money in words

Amount of money in figures

Your name (printed on each cheque)

Your signature

Notice that the money is written in words, in a particular way:

Examples

Thirty-four pounds 45p	**£34·45**
Sixteen pounds only	**£16·00**

Exercise 1

Draw 10 simple copies of the cheque in your book and fill them in using the information given below.

1. Pay J. Hardy £6·86.

2. Pay Helen Gough £23·60.

3. Pay Post Office £80 only.

4. Pay Southern Electricity Board £36·58.

5. Pay Newtown Motorcycle Supplies £560·55.

6. Pay Samantha's Boutique £68·60.

7. Pay Automobile Association £49·55.

8. Pay Cash £50 only.

9. Pay Bigwoods Pools Limited £19·50.

10. Pay Ford Motor Company £8750 only.

11. Pay Gas Board £32·65.

12. Pay Water Authority £35·00.

55 Banking: statements

Remember

At regular intervals, your Bank sends you a statement detailing your money movement. The statement looks like the one below:

Detail	Payments	Receipts	Date	Balance
Balance			9 May	£647.42
Rates 625384657861 DDR	£12.00		11 May	635.42
Building Society	75.00		14 May	560.42
241 DE682 DDR	27.17		20 May	533.25
Smith's Garage 649756385 STO	76.15			457.10
Paid through 918632	16.42		26 May	440.68
Giro system 918633	63.14		28 May	377.54
Bank Giro Credit		£482.16	30 May	859.70
918634	14.24		3 June	845.46
918635	82.24		5 June	763.22
Money paid out 918636	13.18		8 June	750.04
by you 918637	20.00		11 June	730.04
Gas 837465731201 DDR	12.00		12 June	718.04

STO Standing Order   BGC Bank Giro Credit
DDR Direct Debit

Money paid regularly by the bank

The person receiving the money says how much is to be paid and when

Money paid *into* the account

Exercise 1

Use the statement above to answer the following questions:

1. What was the balance of the account on 9 May?

2. How much money was in the account when the statement was sent out?

3. What was the balance on 14 May?

4. What was the balance on 28 May?

5. What was the balance on 8 June?

6. How much was the Bank Giro Credit?

7. Can you guess what the £482·16 was?

8. Why is it in a column of its own?

9. What is the final balance?

10. When the next statement arrives, what will be the amount at the top?

11. How much should the building society receive each year?

12. How much should the garage receive in a year?

13. How much is spent each year on gas, if the same amount is spent each month?

56 Banking: using a statement

Remember

Between receiving statements it is important to keep a record of any transactions that you make so that you are aware of how much money you have in your account. Listed below are the transactions that someone made after receiving a statement on August 18th.

		Credit Balance on August 16	644·84
Aug	18	Paid Gas Bill with 96418231 DDR	10·16
	21	Paid rates to Durham Co with 418263403 DDR	12·42
	25	Drew cash with cheque 641283	25·00
	28	Paid car instalment 14626804 STO	74·53
	30	Received pay Bank Giro Credit	450·00
Sept	3	Drew cash with cheque 641284	25·00
	6	Drew cash with cheque 641285	50·00
	8	Sent cheque to Car Insurance 641286	124·16
	11	Paid telephone bill with cheque 641287	12·66
	18	Drew cash with cheque 641288	84·50
	18	Same as August 18	
	21	Same as August 21	
	28	Same as August 28	
	30	Same as August 30	
Oct.	4	Drew cash with cheque 641289	80·00
	6	Paid in cheque 6101446 from Vernons Pools	500·00

Exercise 1

1. Record these transactions as a bank statement. They should be set out in the same way as the statement on the previous page.

2. What was the balance of the account on (a) August 18th?
   (b) August 31st?
   (c) October 6th?

3. What do you think will be paid on October 28th?

4. How much will be paid off the car in a year?

5. How much pay is received in a year?

6. What is the number of the next cheque to be used?

7. Which bills are paid for monthly?

8. How much did he pay out from his account between 16 August and 30 August?

9a. How much did he pay out from his account in September?
   b. Would you expect him to pay out a similar amount in October?

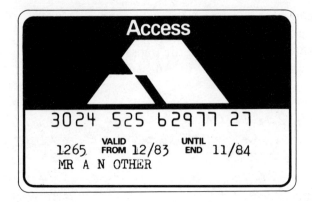
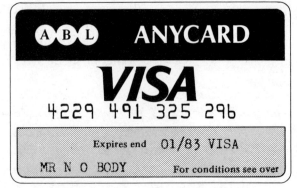

Remember

A credit card is a means of paying for goods that you buy in a shop or through the post. Many people like the convenience of paying one bill at the end of a month, rather than many small bills during the course of the month.

Each transaction that you pay for with your credit card is entered on a statement and sent to you at the end of a month. If you don't pay the full amount that the statement says that you owe, then you will be charged interest on anything that you do owe.

Exercise 1

1. Use the statement below to answer the following questions:

   a. What was paid for on December 12th?

   b. How much interest was charged this month?

   c. How much was owing at the end of last month?

   d. How much was paid off on this statement?

   e. When must some payment reach 'Access' by?

   f. What is the minimum payment?

date	reference number	description	previous balance
			£39·43
		Payment — Thank you	£15·00 —
03 12 81	OP69JD	BRITISH RAIL	£19·20
12 12 81	QR55LB	CANTON RESTAURANT	£12·50
		Interest	£ 1·25
		New Balance	£57·38
			Minimum Payment
Minimum payment should reach us by 30/01/82			£ 7·00

58 Post Office: letters and parcels

	letters			parcels	
weight not over	1st class	2nd class	weight not over	rate
60 g	$15\frac{1}{2}$	$12\frac{1}{2}$	1 kg	£1·20
100 g	22	$16\frac{1}{2}$	2 kg	1·57
150 g	$28\frac{1}{2}$	$20\frac{1}{2}$	3 kg	1·90
200 g	35	26	4 kg	2·10
250 g	42	32	5 kg	2·25
300 g	49	38	6 kg	2·40
350 g	56	44	7 kg	2·55
400 g	64	50	8 kg	2·70

Exercise 1

Calculate:

1. $15\frac{1}{2} \times 3$ 2. 22×5 3. $28\frac{1}{2} \times 6$ 4. $20\frac{1}{2} \times 4$

5. 38×7 6. 44×8 7. £1·20 × 4   8. £2·10 × 3

9. £2·25 × 5   10. £2·55 × 7   11. £2·40 × 9   12. £2·40 × 8

Use the table above to work out the following:

13. Find the cost of:
    a. A letter weighing 24 g first class
    b. A letter weighing 64 g second class
    c. A package weighing 360 g first class
    d. A parcel weighing 1·4 kg
    e. A parcel weighing 3·6 kg.

14. What does it cost to send two letters each weighing 40 g second class?

15. 10 postcards costing 5p each are posted second class. Find the total cost.

16. A mail order firm sends six parcels with weights 2·4 kg, 3·6 kg, 3·8 kg, 4·3 kg, 5·3 kg and 6·5 kg. Find the total cost of postage.

17. Billy Brown bought 10 Christmas cards at 10p each and sent them off by first class post. How much did it cost him?

18. How many first class stamps can be bought for £1·00? How much change will you have?

19. How many letters (all weighing 55 g) can be sent second class for £1·00? How much change will you have?

20. Find the total cost of sending one parcel weighing $2\frac{1}{2}$ kg, one parcel weighing 0·5 kg, 2 letters under 60 g first class, and 3 letters under 60 g second class.

Remember

Using a *postal order* you can name the person who is to receive the money. The Post Office charge *poundage* for making up the postal order: 20p for Postal Orders less than £1·00, 26p for each Order worth £1·00 or more. Postal Orders come in amounts 10p, 15p, 20p and so on to £1. Then £1, £2, £3 and so on to £10.

Postage stamps can be stuck onto a Postal Order to give an amount of money to the nearest penny.

Examples

1. Find the cost of a Postal Order of value 65p.
   Cost = value + poundage
       = 65p + 20p
       = 85p

2. Find the cost of Postal Orders to the value of £6·58.
   Cost = value   + stamp + poundage
       = £6 + 55p + 3p + 26p + 20p
       = £7·04

Exercise 1

Calculate how much the Post Office would charge for the following:

1. A Postal Order of value 40p.

2. A Postal Order of value 69p.

3. A Postal Order of value 46p and one of value 45p.

4. One Postal Order for 96p and another for 47p.

5. Postal Orders to the value of £5·66.

6. Two Postal Orders for 64p each and one for 75p.

7. A Postal Order for £2·20 and another for 65p.

8. Postal Orders to the value of £9·20.

9. Postal Orders to the value of £14·70.

10. Postal Orders to the value of £18·63.

60 Telephones and telegrams

Here is some information on telephone and telegram charges:

| length of call | peak rate | | | standard rate | | | cheap rate | | |
| | mon–fri 9 am–1 pm | | | mon–fri 8 am–9am, 1 pm–6pm | | | all other times | | |
	local	< 56 km	> 56 km	local	< 56 km	> 56 km	local	< 56 km	> 56 km
1 minute	5p	9p	28p	5p	9p	18p	5p	5p	9p
5 minutes	14p	46p	£1.38	9p	32p	92p	5p	14p	32p
10 minutes	23p	92p	£2.76	18p	64p	£1.84	9p	23p	60p

Telegram charges
1. The 'basic' charge is £1.50, and £2.00 for a 'Greetings' telegram.
2. Each word costs 15p, but note that:
   a. each word of the address is included in the charge
   b. a group of figures and words such as P6 is counted as two words; 613AZ1389 counts as three words.

Exercise 1

1. Find the cost of a 1 minute local call on:
   a. Monday 10 am   b. Thursday 2 pm   c. Sunday lunch time

2. Find the cost of a 5 minute call < 56 km on:
   a. Wednesday 8.15 am   b. Friday 5 pm   c. Tuesday 6.15 pm

3. Find the cost of a 10 minute call > 56 km on:
   a. Friday 12.30 pm   b. Friday 1.30 pm   c. Friday 6.30 pm

4. Look at the figures for long-distance calls (> 56 km). *About* how many times cheaper is it if you make your calls at the cheap rate compared with the peak rate?

Exercise 2

Find the cost of these telegrams:

1a. ordinary, 8 words   b. ordinary, 22 words

2a. greetings, 25 words   b. greetings, 30 words

3. *ordinary*: FIRESTONE PRODUCTS CARDIFF WALES STOP PLEASE SEND CLUTCH COVER PART NO 8316AU5 STOP GREENWAYS GARAGE FRONT STREET HULL

4. *greetings*: MR AND MRS SANDERSON 21 EDEN TERRACE BOGNOR STOP HAPPY ANNIVERSARY AND GOOD WISHES FOR FUTURE STOP SORRY WE CANT BE WITH YOU STOP JIM AND JULIE.

5. Write your own telegram congratulating a friend on getting married. Then work out how much it will cost.

61 V.A.T.

Remember

V.A.T. stands for Value Added Tax. It is a tax added on by the government on things you pay for which are regarded as 'luxury items'. A percentage of the price is added as V.A.T. In these examples V.A.T. is 15%.

Example

Find the final cost of a handbag costing £10·00 when V.A.T. is added.

V.A.T. $= 15\%$ of £10·00

$$= \frac{15 \times 10\text{·}00}{100}$$

$$= \frac{150}{100}$$

$$= £1\text{·}50$$

Final price is £10·00 + £1·50

$$= £11\text{·}50$$

Exercise 1

Find the final price on each of the following items:

1. A wristwatch costing £20·00.

2. A telephone bill for £30·00.

3. A sheepskin coat costing £100.

4. A stereo amplifier costing £180.

5. A leather chair costing £250.

6. A double bed costing £300.

7. A motor cycle costing £450.

8. A dining room suite costing £600.

9. A holiday in America costing £1200.

10. Work out the cost of the food in the following bill, find the V.A.T. and calculate the final price.

Bentley Restaurant	£	p
1 minestrone soup		80
3 corns on the cob	2	70
2 T-bone steaks	9	50
1 fillet of plaice	3	75
1 roast duck	4	25
4 lagers	2	40
4 coffees	1	60
TOTAL		
V.A.T.		
FINAL		

62 Calculating V.A.T.

Remember

The following table represents V.A.T. on sums of money between 1p and £100. All of the figures are rounded to the nearest penny.

cost	V.A.T. @ 15%	cost	V.A.T. @ 15%	cost	V.A.T. @ 15%	cost	V.A.T. @ 15%
1p	0p	10p	2p	£1	15p	£10	£ 1·50
2p	0p	20p	3p	£2	30p	£20	£ 3·00
3p	0p	30p	5p	£3	45p	£30	£ 4·50
4p	1p	40p	6p	£4	60p	£40	£ 6·00
5p	1p	50p	8p	£5	75p	£50	£ 7·50
6p	1p	60p	9p	£6	90p	£60	£ 9·00
7p	1p	70p	11p	£7	£1·05	£70	£10·50
8p	1p	80p	12p	£8	£1·20	£80	£12·00
9p	1p	90p	14p	£9	£1·35	£90	£13·50

Example

Find the final price of a radio costing £23·46 when V.A.T. is added.
£23·46 = £20 + £3 + 40p + 6p
from the table V.A.T. = £3·00 + 45p + 6p + 1p = £3·52 (roughly)
final price = £23·46 + £3·52
= **£26·98**

The answer is only approximate because all the figures in the table are rounded to the nearest penny.

Exercise 1

Using the table above, find the V.A.T. on each of the following articles and hence the final price. Your answers will be approximate.

1. A theatre ticket costing £2·40.

2. A meal out for 2 costing £18·30.

3. A portable television costing £74·80.

4. A kettle costing £12·49.

5. An iron costing £22·56.

6. A manual typewriter costing £49·99.

7. A vacuum cleaner costing £65·75.

8. A dinner service costing £89·99.

9. A telephone bill for £46·24.

10. A carpet costing £99·90.

11. A gas bill costing £24·66.

12. A washing machine costing £87·40.

13. A tin of paint costing £8·99.

14. A football costing £15·85.

15. A tracksuit costing £12·62.

63 Car insurance: premiums

Remember

Every car owner must insure his car against accidents. The amount you pay depends on
a) the value of the car
b) the insurance group in which the car is
c) the district in which you live
d) whether you have had other accidents recently
e) your age

This is an example of a table for comprehensive insurance for one year:

Rollover Insurance Company					
insurance group	**district**				
	A	B	C	D	E
1.	£250	£280	£ 312	£ 350	£ 380
2.	£295	£325	£ 360	£ 400	£ 440
3.	£350	£395	£ 440	£ 490	£ 540
4.	£435	£480	£ 535	£ 590	£ 656
5.	£540	£590	£ 650	£ 720	£ 800
6.	£660	£755	£ 853	£ 953	£1050
7.	£860	£950	£1040	£1150	£1280
To each premium add £3 for each £100 of the car's value					

Example

A man living in Inner London (district E) owns a Mercedes-Benz 230 E saloon (insurance group 7). If the value of the car is £10 500, find the cost of his insurance.

Premium = £1280 + (3 × £105)
= £1280 + £315
= **£1595**

Exercise 1

Work out the premium for each car owner for a comprehensive policy from Rollover Insurance:

1. *Mini Metro 1·0 HLE* group 2; district B; value £3700.

2. *Princess 1700 HLS* group 4; district C; value £4300.

3. *Citroen GSA Club* group 4; district D; value £3900.

4. *Colt Lancer 1400* group 6; district A; value £4500.

5. *Ford Escort 1·6 GL* group 5; district E; value £4800.

6. *Rover 2300 S Saloon* group 6; district B; value £6500.

7. *Renault 18 GTL* group 4; district B; value £4900.

8. *Volkswagon Golf LS* group 4; district D; value £4200.

9. *Volvo 265 GLE* group 7; district E; value £9600.

10. *Rolls Royce Silver Spirit* group 7; district E; value £46 800.

64 Car insurance: no-claims bonus

Remember

Car Insurance is very expensive. But if you don't have an accident you receive a reduction on your insurance payment. This is called a *no claims bonus.*

no claim	reduction
1 year	10%
2 years	20%
3 years	50%
4 years	60%

Example

A lady with an Austin Allegro (insurance group 3) lives in district C and her car is worth £2500. She has two years no claims bonus. Find her premium if she insures with Rollover.

Full premium $= £440 + (3 \times 25)$

$= £440 + £75$

$= £515$

No claims bonus $= £515 \times 20\%$

$= £515 \times \frac{20}{100}$

$= £103$

so, premium paid $= £515 - £103$

$= £412$

Exercise 1

Using the 'Rollover' table of premiums from the last page, and the list of no claims bonuses above, calculate the following premiums:

	insurance group	district	value	no claims
1.	1	A	£3000	1 year
2.	6	C	£3900	none
3.	4	B	£6000	1 year
4.	3	D	£7500	3 years
5.	7	E	£6000	none
6.	2	E	£3400	4 years
7.	5	D	£5500	2 years
8.	6	E	£4200	2 years
9.	5	A	£5250	3 years
10.	7	C	£6750	4 years

65 Hire purchase

Remember

Hire purchase is a way of having the use of things before you have finished paying for them. You end up paying more than you would if you pay for them straight away.

Cash Price £65

Deposit £10 only +6 monthly instalments of £12

Example

Find the Hire Purchase price of the bicycle shown above. Find the difference between the hire purchase price and the cash price.

H.P. price = £12·00 instalment

for   6 months

72·00

plus   10·00 deposit

82·00

H.P. charges = H.P. price − Cash price
= £82·00 − £65·00
= **£17·00**

Exercise 1

Find the cost of buying the following items on hire purchase, then calculate how much more this is than the cash price.

1. *Wristwatch* Deposit £2·50 and 6 instalments of £1. Cash price = £7.

2. *Record Cabinet* Deposit £10 and 10 monthly instalments of £5. CP = £45.

3. *Easy Chair* Deposit £6 and 9 monthly instalments of £1·90. CP = £18.

4. *Bookcase* Deposit £16 and 6 monthly instalments of £12·50. CP = £75.

5. *Gas Cooker* Deposit £30 and 12 monthly instalments of £12·50. Cash price is £155.

6. *Indian Carpet* Cash price £350 or £80 deposit and 12 monthly instalments of £28·50.

7. *Caravan* Cash price £2500 or £1000 deposit and 24 monthly instalments of £92

8. *Motor Cycle* Cash price £3200 or £500 deposit and 36 monthly instalments of £113·50.

9. *Washing Machine* Cash price £170 or 10% of cash price as deposit and 12 monthly instalments of £16·40.

10. *Dining room suite* £850 cash price or 20% of cash price as deposit and 24 monthly instalments of £32·60.

66 Rent

Remember   If you don't own the house in which you live, you have to pay *rent* to the owner. All rent paid should be recorded in a *rent book*. Here is a page from Mrs. Green's rent book:

date	rent	rates	arrears	total due	total paid	signature
7th March	£19·20	£2·40	-	£21·60	£21·60	A. Brown
14th March	£19·20	£2·40	-	£21·60	£21·60	A. Brown
21st March	£19·20	£2·40	-	£21·60	-	
28th March	£19·20	£2·40	£21·60	£43·20	-	
4th April	£19·20	£2·40	£43·20	£64·80	£64·80	A. Brown

Exercise 1

1. What was the weekly rent?

2. How much is paid weekly for rates?

3. What is the total amount paid each week for rent and rates?

4. What was the total amount due for the month of March?

5. How much was actually paid in March?

6. The tenant pays rent for 50 weeks and has 2 weeks free. Find how much she pays in a year.

7. How much is paid in rates for a year? (50 weeks)

8. What is the total amount paid each year in rent and rates?

9. What is the name of the rent collector?

10. Suggest one reason why Mrs. Green got behind in her rent.

11. Was Mrs. Green in arrears on 5th April?

12. Make out a table like the one above and fill it in using the following information:
    a. Money is collected on 1st, 8th, 15th, 22nd and 29th December
    b. The weekly rent is £16·80
    c. The rates are £2·64
    d. The collector got no answer on 8th and 22nd December
    e. The collector is you.

13. How much rent should be paid in December?

14. What was the total cost of rates for December?

15. How much was paid in rates in a year (50 weeks)?

16. How much was paid in rent in a year (50 weeks)?

67 House purchase

Remember

Most people borrow money to buy a house. A sum of money borrowed from a building society or a bank to pay for a house, is called a *mortgage*. It is repaid by monthly instalments usually over 20 or 25 years. The table below is a repayment table:

	monthly instalments	
mortgage	over 20 years	over 25 years
£ 8000	£100	£ 88
£ 9000	£112	£ 98
£10 000	£125	£108
£11 000	£138	£119
£12 000	£150	£130
£13 000	£163	£140
£14 000	£175	£151
£15 000	£188	£162
£18 000	£225	£195
£20 000	£250	£216

Examples

1. What would be the monthly repayment on a mortgage of £12 000 over 25 years?
   Monthly payment = **£130** (from the table above)

2. A house costs £10 000. A man puts down 20% deposit and pays off his mortgage in 20 years. Find the value of the deposit and the monthly repayments.

$$\text{Deposit} = 20\% \text{ of } £10\ 000$$
$$= \frac{20}{100} \times 10\ 000$$
$$= £2000$$
$$\text{Mortgage} = £10\ 000 - £2000$$
$$= £8000$$
$$\text{Monthly repayments} = £100$$

Exercise 1

1. Mr. Jackson buys a bungalow and obtains a mortgage of £8000 over 25 years.
   a. How much does he repay each month?
   b. How much does he repay each year?

2. Mrs. Johnson buys a house for £14 000. She pays £3000 deposit and receives a mortgage for the remainder to be repaid over 25 years. Find her monthly payment.

3. Mr. Richardson has a mortgage of £12 000 over a period of 20 years. To buy his house he put down a deposit of £3000.
   a. What was the total price of the house?
   b. What is his monthly repayment?

4. A terraced house costs £12 000. Jimmy Atkinson puts down 25% deposit and receives a mortgage for the remainder over 20 years.
   a. How much was the deposit?
   b. What were his monthly repayments?

5. Olaf Erikson buys a Swedish type bungalow for £20 000. He puts down 25% deposit and obtains a mortgage for the remainder to be repaid over 20 years.
   a. Find the value of his deposit.
   b. What was his monthly repayment?

6. A detached cottage cost £19 000. Colin Clarkson paid a deposit of £4000 and obtained a mortgage for the rest over 25 years.
   a. Find his monthly repayment.
   b. How much does he pay each year?

7. Mr. Michaelson bought a semi-detached house for £15 000. He puts down a deposit of 20% and repays the rest over 20 years.
   a. What was his deposit?
   b. What was his monthly repayment?
   c. How much did he pay in a year?

8. Ronnie Thompson bought a house for £17 000. He paid £2000 deposit and the remainder over 20 years.
   a. What was his monthly repayment?
   b What was his yearly repayment?
   c. How much did he pay altogether over the full 20 years?

9. Mr. Parker bought a bungalow for £25 000. He paid 20% deposit and his mortgage was over 25 years.
   a. What was his deposit?
   b. What was his monthly repayment?
   c. Find the amount paid after one year.
   d. What was the total cost after 25 years?

10. Mr. and Mrs. Gowland bought a public house for £30 000. They paid 50% deposit and the remainder was to be paid over 20 years.
    a. What was their deposit?
    b. What is their monthly repayment?
    c. How much do they pay each year?
    d. Find the total cost after 20 years.

68 Household insurance

Remember

Once you have bought a house, it is a good idea to insure the contents against damage or burglary. Below is an example of a household insurance premium guide. Premiums vary depending on where you live.

package reference letter		A	B	C	D	E	F
contents — sum insured		£1500	£2000	£2500	£3000	£3500	£4000
personal effects and valuables		£ 500	£ 500	£ 750	£ 750	£1000	£1000
premiums	District 1	£12·25	£13·50	£16·25	£17·50	£18·75	£21·75
	District 2	£14·75	£16·25	£19·25	£21·50	£23·00	£26·75
	District 3	£17·30	£19·40	£25·25	£27·35	£29·45	£35·30

Exercise 1

1. Which package would you require for each of the following set of circumstances:
   a. Contents £3000: personal effects £750.
   b. Contents £2500: personal effects £750.
   c. Contents £2000: personal effects £500.
   d. Contents £3500: personal effects £1000.
   e. Contents £4000: personal effects £1000.

2. Find the premium for each of the answers in question **1** if all the houses insured were in district 2.

Find the premiums paid in the following examples:

3. Package A in district 3.

4. Package B in district 1.

5. Package E in district 3.

6. Package D in district 1.

7. Contents £1500: personal effects £500: district 2.

8. Contents £2000: personal effects £500: district 1.

9. Contents £3000: personal effects £750: district 3.

10. Contents £4000: personal effects £1000: district 3.

11. Contents £2500: personal effects £750: district 2.

12. Contents £3500: personal effects £1000: district 1.

13. Contents £2500: personal effects £750: district 3.

14. Contents £4000: personal effects £1000: district 1.

15. Contents £2000: personal effects £500: district 2.

16. Contents £1500: personal effects £500: district 3.

69 Life assurance

Remember

Many people take out life assurance. This means that for a monthly payment your family will receive a certain sum of money when you die. The table below shows a selection of yearly premiums for each £100 of Assurance:

age	premium	age	premium	age	premium
21	£5·00	31	£5·75	41	£7·25
22	£5·04	32	£5·87	42	£7·48
23	£5·08	33	£5·98	43	£7·73
24	£5·15	34	£6·10	44	£7·99
25	£5·22	35	£6·22	45	£8·30
26	£5·29	36	£6·34	46	£8·65
27	£5·37	37	£6·49	47	£9·05
28	£5·45	38	£6·66	48	£9·50
29	£5·55	39	£6·84	49	£10·00
30	£5·65	40	£7·04	50	£10·65

Examples

1. Find the annual premium of a man who wishes his dependents to receive £2000. He is 36 years of age.

Annual Premium $= £6·34 \times \frac{2000}{100}$

$= £6·34 \times 20$

$= £126·80$

2. A woman aged 22 takes out a life assurance policy for £5000.
   a. What is her annual premium?
   b. What is her monthly payment?

Annual premium $= £5·04 \times \frac{5000}{100}$

$= £5·04 \times 50$

$= £252·00$

Monthly payment $= £252·00 \div 12$

$= £21·00$

Exercise 1

Use the above table to work out the following questions:

1. A man aged 27 wishes to take out life assurance for £5000. Find his annual premium.

2. Work out the annual premium of a man who wishes to assure his life for £1000. He is 50 years of age.

3. A woman 34 years of age insures her life for £6000.
   a. What is her annual premium?
   b. Find her monthly payment.

In the following sums find the annual premium then work out how much the monthly repayments would be:

	name	age	amount assured
4.	A. Amberside	28	£2000
5.	B. Blacklaw	30	£4000
6.	G. Greener	36	£6000
7.	P. Pinkney	43	£8000
8.	W. Whitelaw	48	£10 000

70 Electricity

Remember

> Electricity used   = Power rating of appliance × No. of hours used
> (in kilowatt hours, or **kWh**)   (in kilowatts, or **kW**)   (in hours, or **h**)

1 kilowatt = 1000 watts.

Examples

1. How much electricity does a 2kW fire use in 3 hours?
   Electricity used = 2 kW × 3 = **6 kilowatt hours**

2. How much electricity does a 100 watt bulb use in 10 hours?
   Electricity used = $\frac{100}{1000}$ × 10 = **1 kilowatt hour**

Exercise 1

Work out how much electricity is used by each of these appliances:

1. A 3 kW fire used for 3 hours.

2. A 4 kW fire used for 4 hours.

3. A $2\frac{1}{2}$ kW fire used for 6 hours.

4. A 6 kW fire used for $\frac{1}{2}$ hour.

5. A 8 kW fire used for 15 minutes.

6. A 100 watt light bulb shining for 20 hours.

7. A 250 W light bulb shining for 12 hours.

8. A 500 W photographer's light switched on for 2 hours.

Remember

A kilowatt hour is usually called a *unit* by the electricity board. The charge for one unit of electricity is about 6p.

Example

How much does it cost to run a 3 kW fire for 5 hours?
Electricity used = 3 kW × 5 = 15 kilowatt hours or units
     Cost = 6p × 15   = 90p or **£0·90**

Exercise 2

Find the cost of electricity needed to run the following appliances:

1. A 4 kW fire burning for 4 hours.

2. A 5 kW fire burning for 6 hours.

3. A 8 kW immersion heater switched on for $\frac{1}{2}$ hour.

4. A 4 kW suntan lamp switched on for $1\frac{1}{2}$ hours.

5. A 4 kW washing machine used for $4\frac{1}{2}$ hours.

6. A 2 kW deep freeze used for 12 hours.

7. A 500 W bulb switched on for 24 hours.

8. A 100 W bulb left on for 3 full days.

9. A 250 W bulb lit for 16 hours.

10. A $\frac{1}{2}$ kW soldering iron used for 36 hours.

71 Reading the meter

Remember

The electricity meter that you have records how much electricity you use in kilowatt hours or *units*. When an inspector calls to read the meter he records the numbers on the dials and so is able to calculate how much electricity has been used since it was last read.

Example

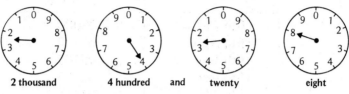

2 thousand   4 hundred   and   twenty   eight

If the pointer is *between* two numbers, the one you want is always the *lower* of the two.

Exercise 1

Read the following meters:

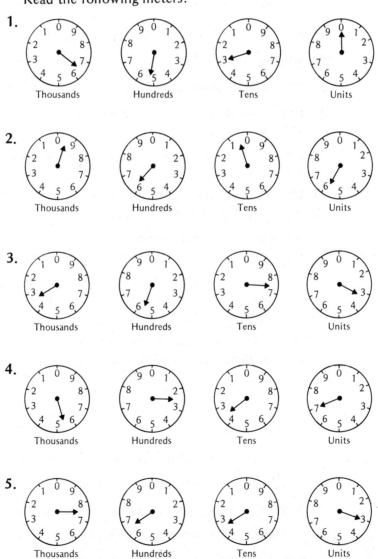

1.   Thousands   Hundreds   Tens   Units

2.   Thousands   Hundreds   Tens   Units

3.   Thousands   Hundreds   Tens   Units

4.   Thousands   Hundreds   Tens   Units

5.   Thousands   Hundreds   Tens   Units

72 Electricity bills

Remember

At regular intervals, an inspector comes to read your meter. Part of a page in his book for your meter could look like this:

date	reading (units)	consumption (units)
9th Jan	00000	00000
16th April	01320	01320
10th July	02105	00785
15th Oct	02715	. . . . .
10th Jan	03718	. . . . .
12th Apr	04936	. . . . .
16th July	05689	. . . . .
12th Oct	06298	. . . . .

Exercise 1

1. Copy the table and complete the 'consumption' column.

2. What was the reading on the meter on 16th April?

3. What was the reading on the meter when it was read on 10th July?

4. How many units of electricity were used between 9th January and 16th April?

5. How many units of electricity were used between 16th April and 10th July?

6. Why do you think that more electricity was used between January and April than between April and July?

7. When was this meter installed?

Remember

You can work out how big your electricity bill is going to be by reading the meter yourself and then using the Electricity Board's rate of charge. The rate is about 6p a unit.

Example

What will be Joe Spark's electricity bill if he uses 01346 units?

Cost of electricity = 01346 × 6p

$$\begin{array}{r} 1346 \\ \times \quad 6p \\ \hline 8076p \\ \hline = £80.76 \end{array}$$

Exercise 2

Work out the cost of the following electricity bills:

1. F.R. Eezing who has used 00450 units in winter.

2. R. Oasting who has used 01200 units in summer.

3. A. Householder who has used 00624 units.

4. R.E. Dhot who has used 02751 units.

5. Miss C. Alculate whose last two meter readings were 02105 and 02834.

6. F. Ridge whose last two meter readings were 04623 and 05378.

7. S. Unlamp whose last two meter readings were 06846 and 07488.

8. B. Ulb whose last two meter readings were 08364 and 09456.

9. B. Attery whose last two meter readings were 04738 and 05688.

10. S.C. Orch whose last two meter readings were 03345 and 04788.

73 Gas bills

Remember

You can pay for the gas you use in different ways. These ways are called tariffs and are scales of charges set out by the Gas Board. Gas bills usually come every three months. Gas is charged for in *therms*.

domestic credit	A fixed charge of First 52 therms @ Remainder @	£6·40 25p each 24p each
general credit	A fixed charge of All therms @	£9·00 30p each
power generation	A fixed charge of All therms @	£2·00 32p each

Example

Find the cost of 160 therms on the Domestic Credit tariff.
Fixed charge =   £6·40
52 therms @ 25p = £13·00
(Remainder) 108 therms @ 24p   = £25·92
Total Cost = **£45·32**

Exercise 1

Calculate the gas bill for each of the following consumers:

1. William Bull using 200 therms on Domestic Credit.

2. I.T. Snatural using 300 therms on Domestic Credit.

3. Gas Pipes Ltd. consuming 1000 therms on Power Generation.

4. 'Itsagas' using 500 therms on General Credit.

5. 'General Supplies' store using 650 therms on General Credit.

6. Leaks Ltd. consuming 2000 therms on Power Generation.

7. Nosmo King using 350 therms on the Domestic Credit.

8. S. Mell using 285 therms on General Credit.

9. D. Entist using 463 therms on General Credit.

10. Blue Flame Ltd. using 1542 therms on Power Generation.

74 Foreign currency

Remember

If you visit a foreign country you must exchange British currency for the currency of the country you visit. The amount of foreign money you obtain depends on the rate of exchange, which varies from week to week. Here is a table showing some typical rates of exchange:

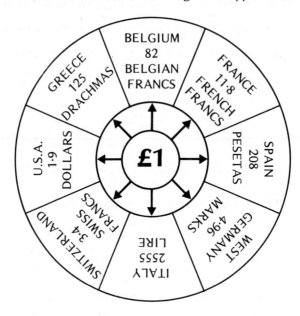

Example

Change £60 into Belgian francs
£60 = 60 × 82 Belgian francs
    = **F. 4920**

Exercise 1

1. Change £10 into French francs.

2. Change £30 into drachmas.

3. Change £40 into dollars.

4. Change £50 into lire.

5. Change £100 into pesetas.

6. Change £120 into marks.

7. A man goes on his holidays to Switzerland for two weeks and takes £150 with him which he changes into Swiss francs. How many francs does he receive?

8. Bill and Ben go on holiday to Spain. Bill changes £90 into pesetas and Ben changes £70. How many more pesetas does Bill have than Ben?

9. Peter and Paul travel to U.S.A. Peter changes £120 and Paul changes £130. How many dollars do they receive altogether?

10. Taters and Mould go on holiday to Italy. They change £260 into lire, then share the money out equally. How much does each person receive?

75 Holidays abroad

Remember

You may want to take a car abroad and tour the continent with a caravan or tent. One of the largest expenses is crossing the English Channel. Below are two tariffs for making the crossing:

Dover to Calais (single)			
boat		**hovercraft**	
car	£36·00	car	£48·00
adult	£8·50	adult	£10·00
child	£4·25	child	£5·00
caravan	£24·50	caravan	£32·50
Return journey twice single fare			

Example

Jimmy White, his wife and two children make a return journey by hovercraft across the channel with their car.
Find the total cost

1 car @ £48·00	=	£48·00
2 adults at £10	=	£20·00
2 children @ £5	=	£10·00
		£78·00

Return fare = £78·00 × 2
= **£156**

Exercise 1

Work out the prices of the following journeys:

1. One adult travels by hovercraft on a single journey.

2. 2 children travel by boat on a single journey.

3. One adult with his car travels by boat on a single journey.

4. 2 adults on a single journey by hovercraft.

5. 4 children on a single journey by hovercraft.

6. Three adults with their car on a single journey by boat.

7. Five monks with their car on a return journey by boat.

8. Mr. and Mrs. Johnson and their two children. Single journey by boat with their car.

9. Mr. and Mrs. Jones on a return journey with a car and a caravan by boat.

10. Mrs. Duffy and her three children on a return journey. They travel by hovercraft.

11. Mr. and Mrs. Snailpace and their four children with their car and caravan on a return journey by boat.

12. Mr. and Mrs. Russell with their ten-year-old daughter travel by boat on a return journey with their car.

76 Booking a holiday

Remember

When you want to travel abroad for a holiday, it is often best to book up through a travel agent. Below is a price table showing costs per person of a holiday in the Rhineland, Germany:

12 nights at hotel (travel by coach)	£360	£380	£400	£420	£380
14 nights at hotel (travel by plane)	£420	£440	£460	£480	£440
	Up to 2nd June	3rd–23rd June	24/6–7/7 29/8–7/9	8 July– 28 Aug	After 8th Sept
Children under 16 years of age: half price Travel Insurance: £6·50 per person					

Example

Find the cost for two adults and one child to go to the Rhineland by plane on 1st July

2 adults @ £460	=	£920·00
1 child @ £230	=	£230·00
3 travel insurances @ £6·50	=	£19·50
Total Cost	=	£1169·50

Exercise 1

Find the total cost of the following holidays:

1. 2 adults by coach from 3rd June.

2. 3 adults by plane from 8th July.

3. 2 children by plane from 24th July.

4. 3 adults and 2 children by coach from 10th June.

5. Mr. and Mrs. Green by coach from 8th July.

6. Mr. and Mrs. Brown and their 12 year old son by plane from 29th August.

7. The four Gray brothers (in their twenties) by coach from 20th September.

8. Mr. and Mrs. Magenta with their 18 year old son and their 12 year old daughter by plane from 9th August.

9. Mr. Scarlett and his three 6 year old triplets by coach from 29th August.

10. Mr. and Mrs. Black with their three children catching the coach on 15th July.

11. Mr. and Mrs. White and their two children by plane from 21st August.

12. Mr. and Mrs. Pink and their four children by coach from 10th June.

77 Time: the calender

Remember   This is the calender for the year 2000!

	January							February							March							April					
S	M	T	W	T	F	S	S	M	T	W	T	F	S	S	M	T	W	T	F	S	S	M	T	W	T	F	S

```
       January                    February                     March                       April
 S  M  T  W  T  F  S         S  M  T  W  T  F  S         S  M  T  W  T  F  S         S  M  T  W  T  F  S
                   1                  1  2  3  4  5                  1  2  3  4                              1
 2  3  4  5  6  7  8         6  7  8  9 10 11 12         5  6  7  8  9 10 11         2  3  4  5  6  7  8
 9 10 11 12 13 14 15        13 14 15 16 17 18 19        12 13 14 15 16 17 18         9 10 11 12 13 14 15
16 17 18 19 20 21 22        20 21 22 23 24 25 26        19 20 21 22 23 24 25        16 17 18 19 20 21 22
23 24 25 26 27 28 29        27 28 29                    26 27 28 29 30 31           23 24 25 26 27 28 29
30 31                                                                               30

         May                        June                        July                       August
 S  M  T  W  T  F  S         S  M  T  W  T  F  S         S  M  T  W  T  F  S         S  M  T  W  T  F  S
    1  2  3  4  5  6                     1  2  3                           1            1  2  3  4  5
 7  8  9 10 11 12 13         4  5  6  7  8  9 10         2  3  4  5  6  7  8         6  7  8  9 10 11 12
14 15 16 17 18 19 20        11 12 13 14 15 16 17         9 10 11 12 13 14 15        13 14 15 16 17 18 19
21 22 23 24 25 26 27        18 19 20 21 22 23 24        16 17 18 19 20 21 22        20 21 22 23 24 25 26
28 29 30 31                 25 26 27 28 29 30           23 24 25 26 27 28 29        27 28 29 30 31
                                                        30 31

       September                   October                    November                    December
 S  M  T  W  T  F  S         S  M  T  W  T  F  S         S  M  T  W  T  F  S         S  M  T  W  T  F  S
                1  2         1  2  3  4  5  6  7               1  2  3  4                           1  2
 3  4  5  6  7  8  9         8  9 10 11 12 13 14         5  6  7  8  9 10 11         3  4  5  6  7  8  9
10 11 12 13 14 15 16        15 16 17 18 19 20 21        12 13 14 15 16 17 18        10 11 12 13 14 15 16
17 18 19 20 21 22 23        22 23 24 25 26 27 28        19 20 21 22 23 24 25        17 18 19 20 21 22 23
24 25 26 27 28 29 30        29 30 31                    26 27 28 29 30              24 25 26 27 28 29 30
                                                                                    31
```

Exercise 1   Use the calender for the year 2000 to answer the following questions:

1. On what day of the week is your birthday in 2000?

2. On what day of the week is the last day in February?

3. On what day of the week is Christmas Day?

4. Is the year 2000 a leap year?

5. What date is the last Monday in the year?

6. How many Tuesdays in the month of May?

7. What date is the first Tuesday in the year?

8. How many Sundays between April 3rd and June 3rd?

9. If you leave for a holiday early on 22nd July and return late on 6th August, how many days (including the dates mentioned) will you have been away?

10. On what day of the week is New Year's Eve?

11. David Warner goes on holiday from 22nd July until 13th August inclusive. How many days is he away?

12. Schools break up for summer on 21st July for six full weeks. On what date do pupils return if they go back to school on a Monday?

78 Time: the clock

Remember   60 seconds = 1 minute; 60 minutes = 1 hour; 24 hours = 1 day.

Examples
1. 3.11 is written as eleven minutes past three.
2. 3.15 is written as quarter past three.
3. 3.30 is written as half past three.
4. 3.42 is written as eighteen minutes to four.
5. 3.45 is written as quarter to four.

Exercise 1   Write each of the following times out in words:

1a. 12.10	b. 7.14	c. 8.20	d. 1.25
2a. 2.30	b. 10.30	c. 9.30	d. 4.30
3a. 11.15	b. 5.15	c. 10.45	d. 6.45
4a. 3.50	b. 2.59	c. 10.35	d. 4.40
5a. 8.31	b. 1.46	c. 5.55	d. 7.36

Exercise 2   Write each of the following times out in figures:

1a. twenty past six	b. sixteen minutes past two
2a. twenty-three minutes past one	b. twenty-five past twelve
3a. half past eleven	b. half past three
4a. quarter past eight	b. quarter past five
5a. quarter to nine	b. quarter to ten
6a. ten to seven	b. twenty to six
7a. twenty-five to four	b. five to twelve
8a. eighteen minutes to eight	b. twelve minutes to two
9a. four minutes to seven	b. twenty-nine minutes to three
10a. quarter to one	b. one minute to one
11a. sixteen minutes to four	b. eleven minutes past five
12a. twenty to ten	b. twenty past five
13a. twenty-five to twelve	b. twenty-nine minutes past one
14a. twenty-nine minutes to two	b. six minutes past eight
15a. six minutes to seven	b. eleven minutes to five
16a. one minute past ten	b. one minute to ten

79 Time: a.m. and p.m.

Remember

The time 7.30 could be the time that you get up in the morning or the time that you watch your favourite TV programme in the evening.

The day is split up into two halves:
between 12 midnight and 12 noon: use **a.m.**
between 12 noon and 12 midnight: use **p.m.**

Examples

1. 8.30 a.m. means 8.30 in the morning.

2. 9.30 p.m. means 9.30 in the evening.

3. 12.30 p.m. means 12.30 in the afternoon — lunch time.

4. 12.30 a.m. means 12.30 in the morning — just after midnight.

Exercise 1

Write down the time in numbers, with a.m. or p.m., for each situation:

1a. 9.00 in the morning   b. 10.20. Before lunch

2a. 1.22 in the afternoon   b. 3.45. Near the end of school

3a. 5.40. The sun is rising b. 8.10. The sun is setting

4a. 10.20. Back from an evening out   b. 3.20. Its very quiet and dark

5a. 7.30. Breakfast time   b. 6.05. The cock crows

6a. 12.20. Lunchtime   b. 12.50. Most people are asleep

Exercise 2

Try to think of an appropriate time for each of the following situations, and then write it down.

1. Waking up hungry during the night.

2. Waking up an hour before the alarm rings.

3. Breakfast.

4. Walking through the school gates.

5. Morning break.

6. Lunchtime.

7. Bell rings to go home. Hurrah!

8. Watch T.V.

9. Doing your homework.

10. Coming home late from a party.

11. Eating your supper.

12. Going to bed.

80 Time: the 24-hour clock

Remember   In airports and railway stations it is now common to see the time in the 24-hour system. All times are taken from midnight:

Examples
1.   7.40 a.m. = **07.40**
2.   10.16 a.m. = **10.16**   **All 24-hour times contain**
3.   1.30 p.m. = **13.30**   **4 figures**
4.   11.57 p.m. = **23.57**
5.   12.03 a.m. = **00.03**

Exercise 1   Write each of the following times in the 24-hour system:

1a. 10.20 a.m.	b. 11.15 a.m.	c. 10.20 p.m.
2a. 4.15 p.m.	b. 4.30 p.m.	c. 11.47 p.m.
3a. 3.15 a.m.	b. 7.55 a.m.	c. 9.15 a.m.
4a. 9.50 p.m.	b. 8.08 p.m.	c. 1.00 a.m.
5a. 11.00 a.m.	b. 5.30 a.m.	c. 1.00 p.m.
6a. 12.30 a.m.	b. 6.45 p.m.	c. 10.50 a.m.
7a. 7.25 p.m.	b. 7.10 p.m.	c. 6.40 p.m.
8a. 8.45 a.m.	b. 12.45 p.m.	c. 9.34 p.m.
9a. 9.48 p.m.	b. 11.16 p.m.	c. 11.45 a.m.
10a. 10.55 a.m.	b. 6.59 a.m.	c. 12.50 a.m.

Exercise 2   In the next exercise the times are written down in the 24-hour system. Write down your answers using the 12-hour system:

1a. 11.58	b. 02.55	c. 04.22
2a. 04.50	b. 05.34	c. 17.50
3a. 03.45	b. 22.50	c. 16.48
4a. 14.40	b. 23.25	c. 03.35
5a. 18.55	b. 08.45	c. 12.35
6a. 09.40	b. 01.30	c. 00.24
7a. 02.50	b. 13.35	c. 00.50
8a. 15.30	b. 15.40	c. 03.55
9a. 19.20	b. 04.25	c. 14.52
10a. 21.45	b. 05.37	c. 01.38

81 Time: subtraction

Remember

When looking at timetables for buses of trains you need to subtract the times to see how long the journey takes. This is easiest done using the 24-hour notation.

Examples

1. A train leaves a station at 12.20 and arrives at its destination at 17.55. Find the time taken for the journey.

    17.55
    12.20−
    ─────
    5.35

    Journey is **5 hours 35 minutes**

2. Find the time taken to lay a concrete path which was started at 9.30 a.m. and finished at 2.45 p.m.

    2.45 p.m. = 14.45
    9.30 a.m. = 09.30−
    ──────
    5.15

    Time taken = **5 hours 15 minutes**

3. You need to be careful when subtracting minutes if the top line is smaller than the bottom line:

    16.20
    13.55−
    ─────
    2.25

4. Finding journey times that span midnight needs a special method:
    Find the time taken by a train which starts at 2.25 p.m. and arrives at 3.35 a.m. the next day.
    First, subtract from midnight:
    24.00
    14.25−
    ─────
    9.35

    Then, add the time after midnight:
    9 hrs 35 mins + 3 hrs 35 mins
    = **13 hours 10 minutes**

Exercise 1

Subtract the following times:

1a. 19.30
    14.20−

b. 16.45
   11.30−

c. 22.50
   18.35−

d. 18.58
   14.33−

2a. 21.30
    17.50−

b. 16.20
   11.25−

c. 14.25
   08.48−

d. 17.12
   11.54−

Work out how many hours and minutes there are between the following:

3a. 8.00 a.m. and 9.15 a.m.

b. 7.20 p.m. and 10.40 p.m.

4a. 7.50 a.m. and 10.55 a.m.

b. 12.40 p.m. and 5.55 p.m.

5a. 1.20 p.m. and 3.50 p.m.

b. 5.05 a.m. and 7.00 a.m.

6a. 3.40 p.m. and 6.20 p.m.

b. 6.45 a.m. and 12 noon

7. 19.30 on Tuesday and 06.20 on Wednesday.

8. 16.44 on Thursday and 03.15 on Friday.

9. 6.30 p.m. on Monday and 8.25 a.m. on Tuesday.

10. 7.45 p.m. on Saturday and 10.05 a.m. on Sunday.

93

Here is part of a railway timetable:

	Portsmouth	Southampton	Winchester	Basingstoke	Paddington
train R	20.30 d.	20.58	21.20	22.24	23.30
train S	22.18 d.	22.48	23.10	00.36	01.26
train T	23.40 d.	00.14	00.36	01.58	02.54

Except for the times marked d, the times shown are times of arrival.

Exercise 1

Use the timetable above to answer the following questions:

1. Which train reaches Paddington before midnight?

2. Which train takes exactly three hours to get from Portsmouth to Paddington?

3. You're meeting your mother in Basingstoke at midnight. Which train would you have to catch? How long would you have to wait in Basingstoke before meeting your mother?

4. If you arrive at Portsmouth at nine-thirty in the evening, how long would you have to wait for a train to Paddington?

5. Between which two stations is train S at midnight?

6. If it takes 1 hour 22 minutes to travel between Winchester and Basingstoke, which train are you on?

Find out how long each of the following journeys takes:

7. Portsmouth to Southampton on train S.

8. Winchester to Paddington on train R.

9. Southampton to Paddington on train T.

10. Winchester to Basingstoke on train T.

11. Winchester to Basingstoke on train R.

12. Portsmouth to Winchester on train T.

13. Portsmouth to Basingstoke on train T.

14. Portsmouth to Basingstoke on train R.

15. Southampton to Paddington on train S.

16. Southampton to Basingstoke on train R.

17. Portsmouth to Paddington on train S.

18. Portsmouth to Winchester on train S.

19. Southampton to Basingstoke on train T.

20. Winchester to Paddington on train T.

83 Using timetables

The longest train route in Britain is from Penzance (near Land's End) to Thurso (near John O' Groats).
Here is a timetable for the route:

distance from Land's End in miles	city	time
0	Penzance (for Land's End)	Depart 16.00
112	Exeter	Depart 17.50
184	Bristol	Depart 19.48
270	Birmingham	Depart 20.32
343	Manchester	Depart 21.50
455	Carlisle	Depart 23.46
549	Glasgow	Depart 01.24
576	Stirling	Depart 02.18
592	Perth	Depart 02.34
706	Inverness	Depart 04.06
849	Thurso	Arrive 06.52

Exercise 1

Use the timetable to answer these questions:

How long does it take to travel from:

1. Land's End to Exeter.
2. Land's End to Birmingham.
3. Stirling to John O' Groats.
4. Exeter to Bristol.
5. Manchester to Carlisle.
6. Perth to Inverness.
7. Birmingham to Perth.
8. Birmingham to John O' Groats.
9. Exeter to Inverness.
10. Manchester to Stirling.

How far is it from:

11. Exeter to Manchester.
12. Manchester to Glasgow.
13. Carlisle to Stirling.
14. Manchester to John O' Groats.
15. Glasgow to Perth.
16. Bristol to Inverness.
17. Exeter to Birmingham.
18. Perth to John O' Groats.
19. Glasgow to Inverness.
20. Bristol to Stirling.

84 Distance, speed and time

Remember

Distance, speed and time are the three things that matter when you are travelling. The triangle below shows at a glance the relationship between:

Distance (D), Speed (S) and Time (T).

Place your thumb over the 'D' if you want the formula for Distance, over the S for Speed and over the T for Time.

This gives you:

$D = S \times T$ or, distance = speed $\times$ time

$S = \dfrac{D}{T}$ or, speed = distance $\div$ time

$T = \dfrac{D}{S}$ or, time = distance $\div$ speed

Example

How long does it take for a train to travel 320 km if it averages 80 km per hour?

$$Time = \frac{D}{S} = \frac{320 \ (km)}{80 \ (km/hr)} = \textbf{4 hrs}$$

Exercise 1

Find the time taken for each journey:

1.	$D = 100$ km	$S = 20$ km/hr	$T = ?$ hrs
2.	$D = 200$ km	$S = 40$ km/hr	$T = ?$ hrs
3.	$D = 500$ km	$S = 50$ km/hr	$T = ?$ hrs
4.	$D = 250$ km	$S = 50$ km/hr	$T = ?$ hrs
5.	$D = 360$ km	$S = 40$ km/hr	$T = ?$ hrs
6.	$D = 180$ km	$S = 45$ km/hr	$T = ?$ hrs
7.	$D = 770$ km	$S = 77$ km/hr	$T = ?$ hrs
8.	$D = 1000$ km	$S = 100$ km/hr	$T = ?$ hrs
9.	$D = 200$ km	$S = 25$ km/hr	$T = ?$ hrs
10.	$D = 140$ km	$S = 35$ km/hr	$T = ?$ hrs

Exercise 2

1. How long does it take a long distance lorry to cover 3000 km at an average speed of 60 km p.h.?

2. A motorcycle averages 40 km p.h. How long does it take to travel 20 km?

3. A racing car averages 200 km p.h. How long does it take to complete a race over a course which is 1000 km long?

4. How long does it take a car to go 300 km at 60 km p.h?

5. How long does it take for a lorry to travel 2000 km at an average speed of 40 km p.h.?

85 Speed, distance and time

Remember
The distance, speed, time triangle on the last page also gives:

$$S = \frac{D}{T} \qquad \text{and} \qquad D = S \times T$$

or, speed = distance ÷ time   or, distance = speed × time

Examples

1. A car travels 50 km in 2 hours. What was its *average speed*?

   use $S = \frac{D}{T}$ (because we want to find S)

   $\text{Speed} = \frac{50}{2} = \textbf{25 km per hour}$

2. A bus travels for two hours at an average speed of 24 km per hour. How far does it go?

   use $D = S \times T$

   (because we want to find D)

   $\text{Distance} = 24 \times 2 = \textbf{48 km}$

Exercise 1

Calculate the value of the ? in each sum.

1.	S = 20 km per hour	T = 5 hrs	D = ? km
2.	S = 30 km per hour	T = 6 hrs	D = ? km
3.	S = 45 km per hour	T = 8 hrs	D = ? km
4.	S = 55 km per hour	T = 10 hrs	D = ? km
5.	S = 70 km per hour	T = 12 hrs	D = ? km
6.	S = 82 km per hour	T = 12 hrs	D = ? km
7.	D = 20 km	T = 2 hrs	S = ? km per hour
8.	D = 40 km	T = 2 hrs	S = ? km per hour
9.	D = 60 km	T = 3 hrs	S = ? km per hour
10.	D = 90 km	T = 3 hrs	S = ? km per hour
11.	D = 120 km	T = 4 hrs	S = ? km per hour
12.	D = 160 km	T = 5 hrs	S = ? km per hour

Exercise 2

1. An athlete runs 20 km in 1 hour without stopping. Find her average speed.

2. A cyclist travels 60 km in 4 hours. Find his average speed.

3. A train travels 400 km in 4 hours. What is its average speed?

4. How far does a gas-filled balloon travel in 6 hours if it averages 16 km per hour?

5. How far does Angabout the tortoise move in 4 hours if he ambles along at 30 metres per hour?

6. How long does it take an athlete to run 20 km at 8 km per hour?

7. A plane flies at an average speed of 1000 km per hour. How long will it take to fly 5500 km?

86 Simple interest

Remember

When you borrow money from a Bank, you are charged extra for the use of it. This charge is called the *interest*, and the sum of money borrowed is called the *principal*. There is a formula for finding this interest:

$$\text{simple interest} = \frac{\text{principal} \times \text{rate of interest} \times \text{time}}{100}$$

$$\text{or } S.I. = \frac{P \times R \times T}{100}$$

Example

Find the simple interest charged on £600 borrowed for 6 years at 9%

$$\text{or } S.I. = \frac{600 \times 9 \times 6}{100} = \textbf{£324}$$

Exercise 1

Find the simple interest on the following sums:

1. $P = £100$	$R = 3\%$	$T = 3$ years
2. $P = £200$	$R = 4\%$	$T = 2$ years
3. $P = £200$	$R = 5\%$	$T = 3$ years
4. $P = £500$	$R = 6\%$	$T = 4$ years
5. $P = £800$	$R = 7\%$	$T = 5$ years
6. $P = £900$	$R = 8\%$	$T = 4$ years
7. $P = £1000$	$R = 10\%$	$T = 5$ years
8. $P = £1500$	$R = 15\%$	$T = 3$ years
9. $P = £2500$	$R = 20\%$	$T = 4$ years
10. $P = £3000$	$R = 25\%$	$T = 6$ years

Remember

When the numbers will not cancel as in the example above, your calculation is a bit more difficult:

Example

Find the simple interest charged on £664 for 3 years at 5%.

$$S.I. = \frac{P \times R \times T}{100}$$

$$= \frac{664 \times 5 \times 3}{100}$$

$$= \frac{9960}{100}$$

$$= \textbf{£99·60}$$

$$
\begin{array}{r}
664 \\
\times \quad 5 \\
\hline
3320 \\
\times \quad 3 \\
\hline
9960 \\
\hline
\end{array}
$$

Exercise 2

Find the simple interest on the following sums:

1. $P = £245$	$R = 4\%$	$T = 2$ years
2. $P = £350$	$R = 6\%$	$T = 3$ years
3. $P = £650$	$R = 5\%$	$T = 4$ years
4. $P = £635$	$R = 4\%$	$T = 5$ years

5. $P = £275$ $R = 5\%$ $T = 4$ years

6. $P = £455$ $R = 6\%$ $T = 5$ years

7. $P = £746$ $R = 7\%$ $T = 3$ years

8. $P = £388$ $R = 10\%$ $T = 4$ years

9. Freddy Brown borrows £755 to buy a motor cycle. He agrees to pay the money back over three years at 15% rate of interest. Find the interest charged.

10. Wilfred White borrows £2955 to buy a new car. He agrees to repay the money over two years at 16%. Find the interest charged.

87 Compound interest

Remember

If you invest some money and every time you receive interest you add it to your investment, then the interest will get larger each time. This is called *compound interest*.

Example

Find the compound interest on £2000 invested for 2 years at 5%.

First year interest $= \dfrac{£2000 \times 5 \times 1}{100}$ ($T = 1$ because this is only for the first year)

$= £100$

Investment is now worth £2000 + £100 = £2100

Second year interest $= \dfrac{£2100 \times 5 \times 1}{100}$

$= £105$

Compound interest $= £100 + £105$

$= £\mathbf{205}$

Exercise 1

Find the compound interest on the following sums, over two years:

1. $P = £1000$ $R = 5\%$

2. $P = £2000$ $R = 6\%$

3. $P = £3000$ $R = 7\%$

4. $P = £5000$ $R = 8\%$

5. $P = £7000$ $R = 4\%$

6. $P = £8000$ $R = 9\%$

7. $P = £9000$ $R = 3\%$

8. $P = £10\,000$ $R = 10\%$

9. Joe Sparks who invests £15 000 for two years at a rate of 12%.

10. A. Baker who borrows £20 000 for two years at a rate of 15%.

88 Profit and loss

Remember

If you sell something, you may sell it for more than the price that you paid for it: you are selling at a *profit*. If you sell it for less you are selling at a *loss*. You can express this profit or loss as a percentage of the original cost.

Examples

1. A man bought a caravan for £800 and sold it two years later for £1000. Find his profit as a percentage of the cost price.

Selling price   = £1000
Cost price   =   £800
so, cash profit =   £200

so, profit as a fraction $= £\frac{200}{800} = \frac{2}{8} = \frac{1}{4} = 25\%$

so, **percentage profit   = 25%**

2. After buying a car for £100, Sheila Brown sells it a year later for £800. Find her loss as a percentage of the cost price.

Cost price   = £1000
Selling price =   £800
so, cash loss =   £200

so, loss as a fraction $= £\frac{200}{1000} = \frac{20}{100} = 20\%$

so, **percentage loss   = 20%**

Exercise 1

Find the profit or loss as a percentage of the cost price in each case.

1. A shopkeeper buys eggs at 5p each and sells them at 6p each.

2. A car salesman buys a car for £500 and then sells it for £750.

3. Gary Reaser bought his motorcycle for £800. After three accidents he sold it for £200.

4. A housewife bought a washing machine which cost £150. She later sold it for £75.

5. Brian Black sold his house for £15 000 having bought it a few years previously for £12 000.

6. A coal merchant buying coal at £3·50 a sack and then selling it for £4·20 a sack.

7. An antique dealer bought an old Grandfather Clock for £300. He sold it for £600.

8. Gerry Fashion bought a pair of trousers for £12. He soon got tired of them and gave them away to a charity sale.

9. Richard Larson bought a car for £1000 and later sold it for £800.

10. Michael Moat sold his old suit for £12. He had bought it for £20.

89 Budgeting

Remember

When you receive your net wage you must then decide what money goes where. This is called *budgeting.* Here is a budget for a person who brings home £82·00 per week:

food	£12·00
rent	£16·00
clothes	£14·00
fuel	£15·00
pocket money	£ 8·00
car	£10·00
savings	£ 7·00

Exercise 1

1. Make up a budget for yourself, assuming that you have a net wage of £75·00 per week. Put down the items that you feel to be important.

2. Make up a budget for an average housewife who receives a net wage from her husband of £124·50.

3. Make up a budget for a joiner who has to travel 20 miles to work, lives on his own, and is a heavy drinker. His net wage is £83·00.

4. Make up a budget for a dentist whose net wage per week is £150·60.

5. Make up a budget for a cobbler who has a net wage of £88·00.

Make up budgets for the following people:

6. A steelworker with a net wage of £112·60.

7. A plumber who is also a very keen golfer and brings home £98·60.

8. A dustman who brings home £86·50 and who smokes 60 cigarettes per day.

9. A postman who has a net wage of £110 and bets a lot on horses.

10. A circus clown who has a net wage of £75·60, does not pay rent but has six children.

90 Petty cash accounts

Remember A *petty cash account* is a record of the small amounts of money spent out in an office.

Example Jean kept the petty cash account for her firm. At the beginning of each week she made certain that it contained £20:

		£	p			£	p
June 3	Opening balance	20	00	June 3	Stationery	1	62
					Fares	2	41
				4	Postage		96
				5	Gas meter	3	50
				6	Telephone call		36
				7	Tea fund		86
				7	Milk bill	2	45
				9	Newspapers		78
					Total	12	94
June 10	Balance	7	06		Balance	7	06
	Cash received	12	94				
	Carried forward	20	00				

Exercise 1 Copy the above table, then continue the petty cash account for a second, a third, and a fourth week using the following information:

Second week	June 10	Stationery 68p; fares £1·06
	June 11	Tea fund 65p; gas £1·36
	June 12	Newspapers 86p; cleaner £3·08
	June 13	Typewriter ribbon 65p; telegram £2·77
	June 14	Milk 68p; postage £1·34
Third week	June 17	Postage 56p; gas £2·12
	June 18	Window cleaner £2·50; tea fund 35p
	June 19	Fares 98p
	June 20	Telegram £4·88; newspapers 78p
	June 21	Milk 72p
Fourth week	June 24	Tea fund 86p; telephone call 28p
	June 25	Fares £1·60; birthday card 28p
	June 26	Postage 68p
	June 27	Telegram £1·60; newspapers 86p
	June 28	Milk 90p; chocolates £2·24

91 Running balances

Remember

It is often useful to know the financial state of a business on a given date. A *running balance* quickly gives this figure.

An example of such a balance sheet is shown below. It is incomplete.

date	detail	receipts	expenditure	balance to date
1.8.84	Brought forward			£664·34
4.8.84	Cash sales	£62·41		£726·75
8.8.84	Purchases		£50·62	£676·13
10.8.84	Rent		£30·66	
12.8.84	Sales	£32·44		
	Wages		£124·42	
14.8.84	Gas bill		£15·60	
	Postage		£8·50	
	Sales	£26·60		
	Telephone bill		£18·00	
16.8.84	Sales	£15·65		
	Stationery		£6·48	

Notes:

(a) Any *receipt* is *added* to the balance.

(b) Any *expenditure* is *subtracted* from the balance.

Exercise 1

1. Copy the balance sheet above and then complete the column 'balance to date'.

2. What was the balance at the end of the day's trading on August 12th?

3. Would this firm have been able to afford a new piece of machinery that was offered to them at great discount price of £580 on 15th August?

Exercise 2

Make out a balance sheet in exactly the same way as the one above is made out, for the following transactions:

15.3.82	Balance brought forward £256·00
17.3.82	Cash sales £8·50; telephone bill £17·84
19.3.82	Rent £49·90; purchases £22·64
24.3.82	Sales £14·62; wages £106·42
25.3.82	Gas bill £32·62; fares £4·70
26.3.82	Postage £3·86; sales £138·60
29.3.82	Insurance £3·64; sales £38·66
31.3.82	Wages £92·50
1.4.82	Electricity bill £48·60; sales £21·40
2.4.82	Sales £42·16; purchases £18·60
6.4.82	Postage £3·20; sales £125·60
7.4.82	Wages £83·20
9.4.82	Cheque from M. Duffy £68·60
13.4.82	Sales £24·80

92 Cash accounts

Remember

Anyone who runs a business must by law keep an account of all transactions that are carried out. Balancing the books is very important. In the following example of a cash account, all incoming money, or *receipts* is written on the left. Outgoing money, or *expenditure*, is on the right.

	receipts	£	p		expenditure	£	p
June 18	Balance	22	10	June 18	Postage	10	45
19	B. Johnson	34	50	19	Purchases	5	60
	Sales	16	60		Gas	16	40
20	Sales	41	42	20	Electricity	12	62
21	R. Green	30	36	21	Wages	63	70
	Sales	15	76		Rent	24	50
22	Sales	14	84	22	Purchases	16	55
23	W. White	10	52		Total expenditure	149	82
					Balance taken forward	36	28
	Total income	186	10			186	10

Total income = total expenditure

Make out cash accounts for the following transactions, as in the example:

Exercise 1

May 18	Balance	£ 20·00		
19	Sold goods for cash	£ 40·38	Enter as:	Sales
20	Bought stamps	£ 3·50	,,	Postage
	Purchased goods	£111·44	,,	Purchases
21	Paid telephone bill	£ 32·75	,,	Telephone
22	B. Gough paid cheque	£950·00	,,	B. Gough
	Sold goods for cash	£ 67·45		
23	Paid wages	£339·50	,,	Wages
24	H. Brook paid cheque	£ 34·86		
	Sold goods for cash	£ 54·90		
25	Paid gas bill	£ 56·88	,,	Gas
26	Sold old bacon slicer	£ 23·50	,,	Bacon slicer

Exercise 2	Jan 11	Balance	156·04
	12	Purchased goods	24·16
	12	Paid J. Wood	10·72
	13	Paid electricity bill	8·64
	13	Sold goods for cash	36·75
	14	Paid K. Little	6·65
	15	Purchased goods	27·58
	15	Bought stamps	34·66
	16	Paid wages	88·39
	16	Sold goods for cash	83·50
	17	Bought goods	14·68
Exercise 3	March 9	Balance	754·50
	10	Sold goods for cash	24·68
	10	Paid gas bill	12·86
	11	Bought stationery	5·66
	12	Paid S. Cook	43·70
	13	Sold goods for cash	120·44
	13	Purchased goods for cash	28·55
	14	Paid electricity bill	23·50
	15	Paid insurance	12·40
	16	Paid wages	94·00
	17	Received cheque from D. Dent	56·20
	18	Received cheque from R. Rand	95·50
Exercise 4	May 1	Balance	476·90
	2	Bought stationery	4·86
	3	Purchases	28·55
	3	Paid T. Totten	32·44
	4	B. Atkinson paid his account	74·50
	5	Sold goods for cash	85·66
	6	Sold goods for cash	23·44
	6	Received cheque from A. Adler	64·00
	7	Paid rates	45·00
	8	Paid rent	64·00
	8	Sold goods for cash	39·57
	9	Bought new till	84·60
	9	Received cheque for goods	46·88
	10	Paid wages	84·55
	10	Sold goods for cash	102·44

This crossword is based on the work done in the 'maths for living' section. Copy the crossword below and complete it.

1			2	3		4	5
		6					
	7		8			9	
10		11			12		13
14					15		
			16	17			
18		19					20
21			22			23	

ACROSS

1. If normal rate of pay is 60p per hour, give the overtime rate for time and a quarter.

2. An electrician works 38 hours @ £2 per hour. What is his wage for the week?

4. A driller receives 3p for each base plate that he makes. How much does he receive if he makes 400? [Ans. in £].

6. A teacher receives £460 per month. What is his salary for a year?

8. If double time is paid at £1·20 per hour, what is the normal rate of pay? [Ans. in pence].

10. Find the cost of 4 loaves of bread at 29p each.

DOWN

1. A train covers 600 km in 8 hours. What is its average speed?

2. A plumber earns £23·80 gross per day, but takes home £16·24. How much has been deducted?

3. A turner earns 2p for each object that he makes. If his gross wage is £12·40, how many has he made?

5. Find the total simple interest earned on £200 left in the bank for 3 years at an annual rate of 4%.

7. Write down 9.20 pm, using the 24-hour clock system.

9. Find the cost of 7 books at £2·10 each.

10. How many 50p pieces would you need to make up a total of £5·50?

ACROSS

12. How far does a car travel in 8 hours, with an average speed of 93 km/hour?

14. How much would you pay for 3 glasses of whisky at 40p each?

15. A car travels for 6 hours at an average speed of 45 mph. How far has it moved?

16. A cyclist covers 70 miles in 5 hours. What is his average speed?

19. Write down 'quarter past six in the evening', using the 24 hour clock.

21. Find the cost of 6 eggs if they are priced at 58p per dozen.

22. An electrician receives £104 for a wiring job. If his rate of pay is £4/hour, how long did he take to complete the wiring?

23. A cyclist rides 75 km in 5 hours. Find her average speed.

DOWN

11. How many hours pass between noon on Sunday and midnight on the next Tuesday?

12. Find the cost of 18 postcards @ 4p.

13. Find the simple interest on £500 left in a bank for 1 year at 8% p.a.

16. A boy shared £9·10 equally amongst 5 people. How much did each receive?

17. A petrol attendant works for 13 hours at 32p per hour. What was his wage? [Ans. in pence].

18. An antique dealer bought a set of 6 chairs for £312. What was the cost of each chair?

20. Find the cost of 1125 units of electricity if it is charged @ 4p per unit. [Ans. in £].

Maths for living: **Revision test**

1. Express each of the following sums of money in pence:
   - a. £2·20
   - b. £3·31
   - c. £12·50
   - d. £2·07
   - e. £0·10
   - f. £0·06
   - g. £3
   - h. £20

2. Express each of the following sums of money in pounds:
   - a. 122p
   - b. 306p
   - c. 280p
   - d. 28p
   - e. 2800p
   - f. 2080p
   - g. 2p
   - h. 10p

3a. £1·80
    +£1·78

 b. £2·27
    +£1·74

 c. £13·48
    +£ 8·14

 d. £17·68
    +£ 2·80

4a. £1·28
    −£1·17

 b. £3·75
    −£2·28

 c. £6·50
    −£2·45

 d. £13·42
    −£ 0·54

5a. £5·13
    ×   3

 b. £2·23
    ×   5

 c. £4·56
    ×   4

 d. £3·06
    ×   9

6a. £4·72 ÷ 4

 b. £4·71 ÷ 3

 c. £12·15 ÷ 3

 d. £12·15 ÷ 9

7. 6 eggs cost 36p. How much would the eggs in a two egg omelette cost?

8. A supermarket has a special offer of 'three cans of sweet corn for 72p. How much would it cost for you to buy two cans at the same rate?

9. A baker works for 40 hours in a week and receives 195p per hour. Find his gross weekly wage.

10. Mr. Scott is paid 30p for each sheet of metal which he hammers into shape. In one day he got through 70 sheets. How much was he paid for that day?

11. A history teacher has an annual salary of £5040. Find his gross monthly pay.

12. Assuming that V.A.T. is charged at the rate of 15%, find the final cost when V.A.T. has been added to a bill of £35·00.

13. A washing machine is advertised at a price of £220. It is possible to buy the machine on HP with a £50 deposit and 12 monthly instalments of £20. How much more does the machine cost on HP than the cash price?

14. Matt Duffy borrows £3400 to buy a new car. He agrees to pay the money back over 3 years. Find the simple interest charged if the rate is 15% per annum.

15. Write down the following times with a.m. or p.m.:
    - a. 10.00 in the morning
    - b. 10.00 in the evening
    - c. 8.20 and time to go to school
    - d. 3.00, and few are awake
    - e. 11.30 − TV is closing down
    - f. 12.30 − eat your school lunch

16. Write each of the following times in 24-hour notation:

    a. 7.40 p.m.   **b.** 7.40 a.m.   **c.** 11.20 p.m.   **d.** 11 a.m.
    e. 12.17 p.m.   **f.** 12.17 a.m.   **g.** 8.12 a.m.   **h.** 2.00 a.m.

17. A car travels at an average speed of 30 km/hr for 2 hours and then at 60 km/hr for 1 hour.

    a. How far has it travelled altogether?
    b. How long has it been travelling?
    c. Find the average speed for the full journey.

18. Work out how many hours and minutes there are between the following times:

    a. 8.30 a.m. and 11.15 a.m.
    b. 6.15 p.m. and 9.42 p.m.
    c. 14.20 and 20.46
    d. 9.45 p.m. on Saturday and 8.10 a.m. on Sunday.

19. A train leaves Kings' Cross station at 8.42 a.m. and reaches Newcastle at 11.18 a.m. How long did the journey take?

20.

train	Durham	York
X	09.16	10.08
Y	21.44	22.42

    The timetable shows the times two trains are at the stations named. Which is the quicker train, and by how many minutes?

21. Find the missing distance, speed or time in each of the following:

    a. $D = 140$ km $S = 20$ km per hour $T = ?$ hours
    b. $D = 170$ km $S = ?$ km per hour $T = 2$ hours
    c. $D = ?$ km $S = 80$ km per hour $T = 6$ hours

22. Find the compound interest paid on the following sums, over two years:

    a. $P = £4000$ $R = 5\%$
    b. $P = £6000$ $R = 10\%$
    c. $P = £6000$ $R = 15\%$

23. Find the profit or loss as a percentage of the cost price in each of these:

    a. A car bought for £5000 and sold for £4000.
    b. A shopkeeper buys some cheap yoghurt for £2·00 per litre. No-one wants to buy it so he has to sell it for £1·50 per litre.
    c. An antique dealer buys a painting at an auction for £50 and sells it for £250.

94 Measuring

Remember

You can measure lines on a page using a ruler. They will be measured in *centimetres* (cm) or millimetres (mm). If you are measuring longer distances, such as the length of a classroom or corridor then you should use a metre rule or a tape measure. You would measure them in *metres* (m).

10 mm = 1 cm, 100 cm = 1 m, 1000 mm = 1 m

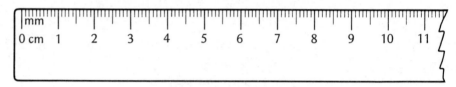

Example

Measure each line, *between the marks*, both in centimetres and millimetres.

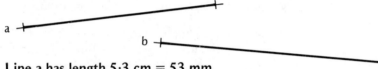

Line a has length 5·3 cm = 53 mm
Line b has length 6·8 cm = 68 mm

Exercise 1

Measure each of the following lines, *between the marks*, using a ruler. Give the answer in centimetres and millimetres:

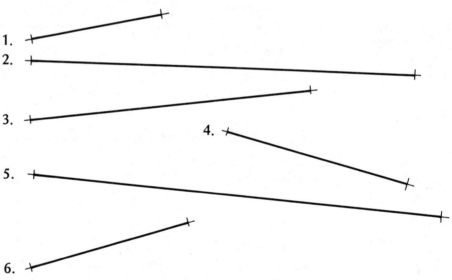

Exercise 2

Using a ruler to draw straight lines, mark off the following lengths:

1a. 12 cm	b. 12 mm	c. 8 cm	d. 70 mm
2a. 16 mm	b. 10 cm	c. 3·2 cm	d. 75 mm
3a. 10 mm	b. 4·3 cm	c. 7·5 cm	d. 100 mm
4a. 5·2 cm	b. 2·9 cm	c. 54 mm	d. 1 cm

95 Estimating

Remember

Sometimes it is useful to be able to estimate how long a line is — perhaps you don't have a ruler, or you're in a hurry. It is impossible to estimate very accurately. The best that you can do with a short line is to estimate it to the nearest centimetre.

Example

Estimate the length of this line and then measure it accurately to find how good the estimate was.

Estimate: 4 cm
Measured length: 6·3 cm
so the estimate was not very good.

Exercise 1

Estimate, to the nearest centimetre, the length of each of the following lines. Write down your estimate and then measure the line with a ruler to see how accurate you were.

1.

2.

3.

4.

5.

6.

7.

8.

9.

10.

96 Construction: lines and triangles

Remember

Lines are parallel if they run in the same direction.

these lines are parallel   these lines are *not* parallel

Example

Draw two parallel lines, 1·5 cm apart.

Method

1. Draw a first line.
2. Mark any two positions on it.
3. Set your compass to 1·5 cm. Draw short arcs.
4. Draw in a line that *just* touches the two arcs.

1.5 cm

Exercise 1

1. Construct two parallel lines, 2·0 cm apart.
2. Construct four parallel lines, with a space of 1·3 cm between each of them.
3. Construct two parallel lines 2·5 cm apart. Make a *parallelogram* by constructing two more lines that cut across them.
4. Make three more parallelograms of any size you like.

Remember

Using a compass and a ruler you can construct triangles of any size.

Example

Construct a triangle with sides of length 3 cm, 4·3 cm and 5 cm.

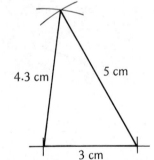

4.3 cm   5 cm

3 cm

Method:
1. Draw a line 3 cm long.
2. From your ruler, set the compass to a radius of 4·3 cm.
3. Put the compass point at one end of the line and draw an arc as shown.
4. Set the compass to a radius of 5 cm.
5. Put the compass point at the other end of the line and draw another arc so that it crosses the first.
6. Complete the triangle by drawing in the lines.

Exercise 2

Construct triangles with these side lengths:

1. 4·8 cm, 4·5 cm and 3·6 cm.
2. 5·6 cm, 6·2 cm and 6·8 cm.
3. 5·8 cm, 6·4 cm and 7·1 cm.
4. 43 mm, 53 mm and 50 mm.
5. 36 mm, 46 mm and 20 mm.
6. 64 mm, 58 mm and 49 mm.
7. 6·2 cm, 68 mm and 5·3 cm.
8. 4 cm, 5 cm and 3 cm (a special triangle)
9. 30 mm, 40 mm and 50 mm.
10. 64 mm, 64 mm, 64 mm (another special triangle).

97 Angles

Remember   A **full turn** = 360°   A **half turn** = 180°   A **quarter turn** = 90°

360°

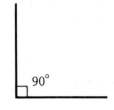

180°

90°

This is also called
a *right angle*.

Examples

1. What angle have you turned through when you have turned through
   2 right angles?
   1 right angle = 90°
   so, **2 right angles** = 2 × 90° = **180°**

2. Through what angle have you turned when you have turned a half
   turn and then back a quarter turn?
   1 half turn = 180°, and 1 quarter turn = 90°
   so, **a half turn and then back a quarter turn** = 180° − 90° = **90°**

Exercise 1   Work out the angle turned through in each case, in degrees.

1a. A half turn   b. A right angle.

2a. A quarter turn   b. A full turn.

3. Two half turns in the same direction.

4. A half turn followed by a quarter turn in the same direction.

5. A full turn followed by another full turn in the same direction.

6. A full turn in one direction, and then a half turn back again.

7. A half turn in one direction, and then a right angled turn in the
   opposite direction.

8. Two right angled turns in one direction followed by a half turn in the
   same direction.

9. A half turn followed by two quarter turns in the opposite direction.

10. A quarter turn, followed by another, then by another, all in the
    same direction.

11. A half turn, followed by a quarter turn in the opposite direction,
    followed by a half turn in the original direction.

12. A quarter turn, followed by a half turn in the same direction, followed
    by a right angled turn in the opposite direction.

113

98 Angle measure

Remember
1. ∠ABC represents the angle that is formed when you run your finger from A to B to C.
2. The place where the angle is, is given by the middle letter.
3. You can only hope to measure an angle to the nearest degree with a protractor. Even that is optimistic!

Example   Measure the following angles using a protractor:

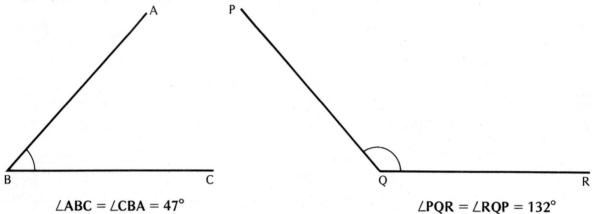

∠ABC = ∠CBA = 47°   ∠PQR = ∠RQP = 132°

Exercise 1   Use a protractor to measure each of the following angles. There are two which you should not need to measure.

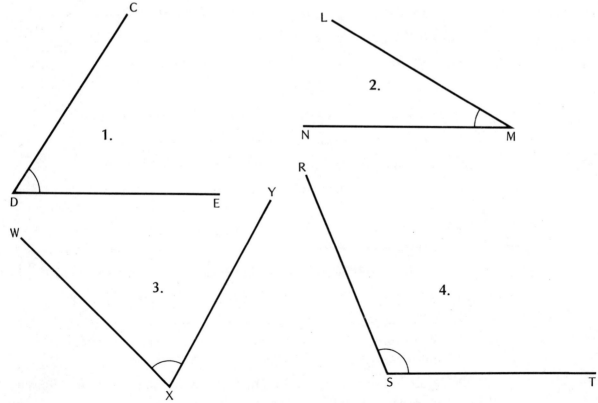

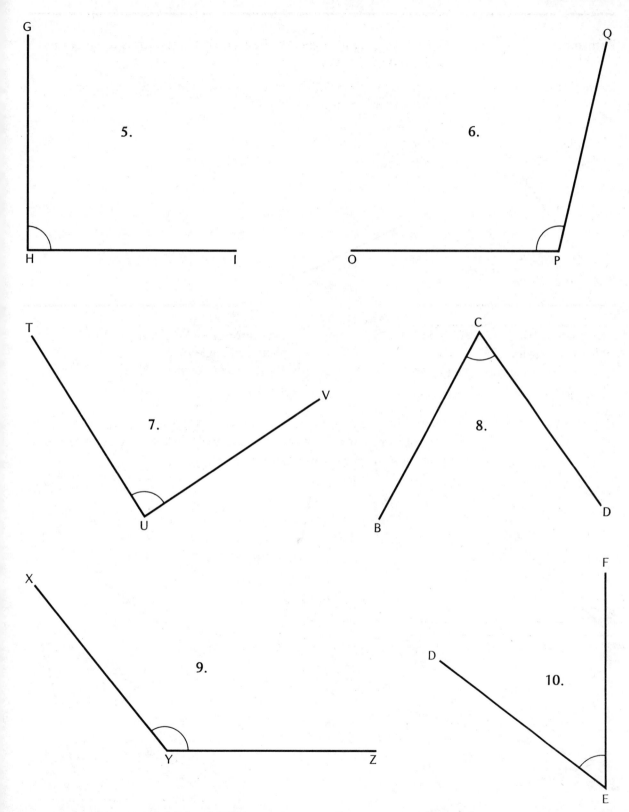

5.

6.

7.

8.

9.

10.

99 Angle estimation and measurement

Remember
For an *estimate* of an angle, the best you can hope for is to be out by no more than 5 degrees.

Example
Estimate each of the following angles and then check your answer by measuring them with your protractor:

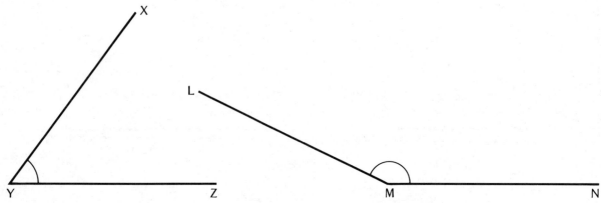

Estimate: $\angle XYZ = \angle ZYX = 55°$
Measured angle: $53°$
So, our estimate was a very good one

Estimate: $\angle LMN = \angle NML = 130°$
Measured angle: $155°$
So, the estimate was not very good

Exercise 1
Estimate the size of each of the following angles in degrees. Write down your estimate. Then measure the angle with a protractor to see how accurate you were with your estimate.

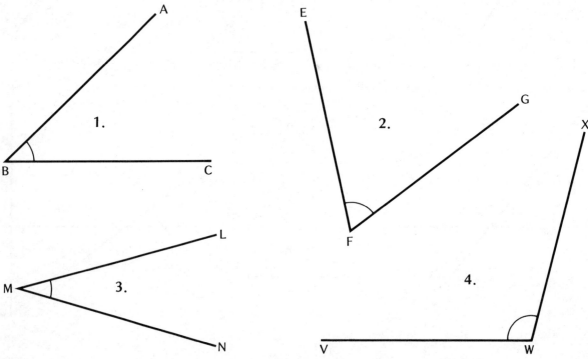

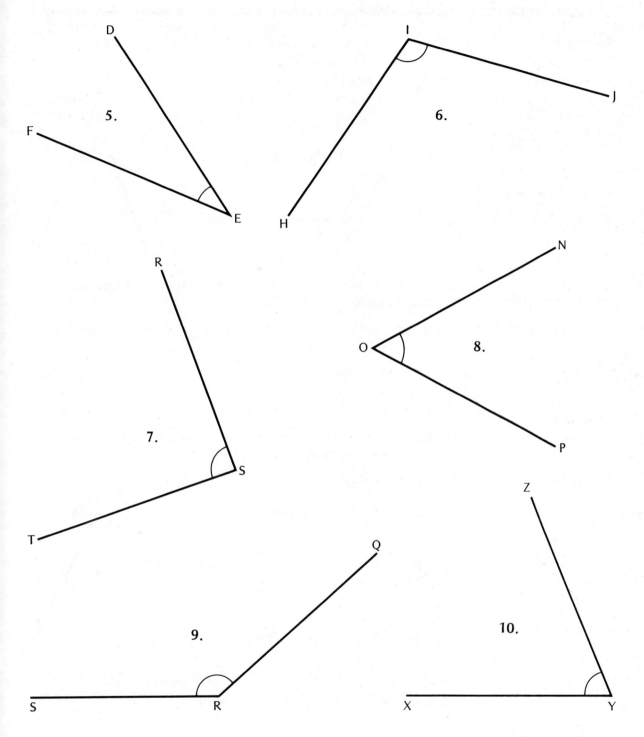

100 Angles: acute, obtuse and reflex

Remember
An *acute* angle
is less than 90°

Examples

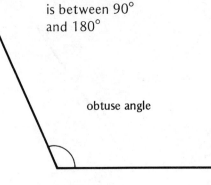

acute angle

An *obtuse* angle
is between 90°
and 180°

obtuse angle

A *reflex* angle
is between 180°
and 360°

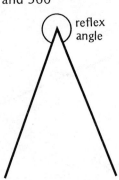

reflex
angle

Exercise 1

1. Go through the exercise in section 98 and identify which questions have acute angles.

2. Go through the exercise in section 99 and identify which questions have obtuse angles.

3. Draw any acute angle.

4. Draw any reflex angle.

Remember

The easiest way to measure the size of a reflex angle is to measure the smaller angle and subtract it from 360°.

Example

Find the marked angles:

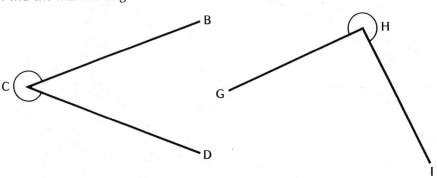

Small (acute) angle = 40°
so, **reflex ∠BCD** = 360° − 40°
= **320°**

Small (obtuse) angle = 93°
so, **reflex ∠GHI** = 360° − 93°
= **267°**

Exercise 2   Find the size of each of the following reflex angles in degrees:

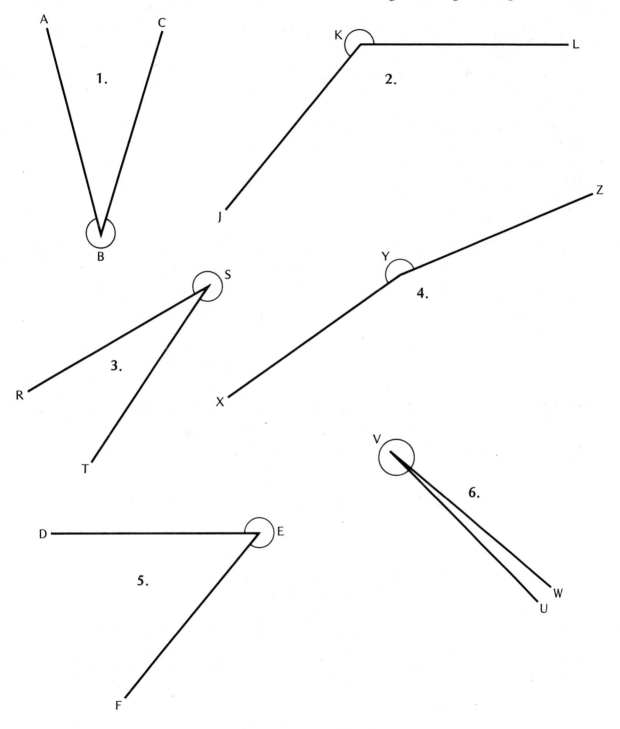

101 Constructing angles

Remember
　　This is how to draw an angle of 67° using a protractor:
1. Draw a base line of any length. Mark a point near one end.
2. Place your protractor at the point, as though you were measuring an angle at that point.
3. 'Measure' an angle of 67°. Put a mark where a line would have to go for there to be an angle of 67°.
4. Draw the line between your new mark and the first one.
5. Mark and label the angle that you have drawn.
6. Measure the angle as a check.

Example
Draw an acute angle of 67°.

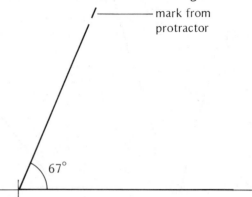

Exercise 1　　Draw each of the following angles in the same way as the example. Some are acute angles whilst others are obtuse.

1a. 40°	b. 60°	c. 30°	d. 80°
2a. 120°	b. 140°	c. 170°	d. 160°
3a. 90°	b. 180°	c. 73°	d. 64°
4a. 163°	b. 93°	c. 89°	d. 179°

Remember　　The easiest way to draw a reflex angle is to draw its related angle and then mark in the reflex angle — in a similar way to the measurement of a reflex angle.

Example　　Draw a reflex angle of 320°.
The related angle is $360° - 320° = 40°$
so, draw an angle of 40° and mark and label the reflex angle.

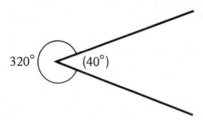

Exercise 2　　Draw each of the following angles in the same way as the example above. All the angles are reflex.

1a. 190°	b. 205°	c. 225°	d. 270°	e. 300°
2a. 330°	b. 216°	c. 229°	d. 184°	e. 343°
3a. 200°	b. 210°	c. 250°	d. 290°	e. 325°
4a. 350°	b. 263°	c. 299°	d. 193°	e. 359°

102 Constructing triangles

Remember

In section 96 you constructed triangles knowing all three sides. There are two other ways.

Example

Construct a triangle with two of the sides being of length 3 cm and 2·6 cm and the angle between them being 60°.

Method

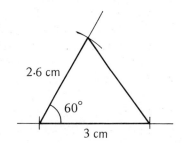

1. Draw a line 3 cm long.
2. At one end measure an angle of 60° and draw the line faintly.
3. On this line mark off a length of 2·6 cm.
4. Join this mark to the other end of the first line.

Exercise 1

Construct triangles with the following measurements:

1. 2·4 cm, 3·4 cm, 30°
2. 3·6 cm, 1·8 cm, 45°
3. 4·2 cm, 2·8 cm, 90°
4. 5·6 cm, 4·1 cm, 70°
5. 2·1 cm, 3·9 cm, 85°
6. 3·2 cm, 4·6 cm, 65°
7. 1·7 cm, 5·1 cm, 55°
8. 3·7 cm, 4·3 cm, 40°
9. 6·4 cm, 3·5 cm, 26°
10. 5·1 cm, 2·6 cm, 22°

Example

Construct a triangle with a side of 5·4 cm which makes angles of 45° and 65° with the other two sides.

Method

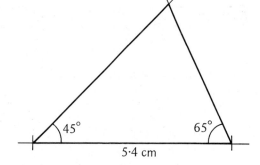

1. Draw a line 5·4 cm.
2. At one end draw an angle of 45° with your protractor.
3. At the other end draw an angle of 65°.
4. Where these lines meet gives the top of the triangle.

Exercise 2

Construct triangles with the following measurements:

1. 6·2 cm, 60°, 40°
2. 4·6 cm, 75°, 36°
3. 38 mm, 22°, 18°
4. 53 mm, 80°, 41°
5. 32 mm, 63°, 36°
6. 4·1 cm, 48°, 61°
7. 66 mm, 63°, 36°
8. 2·7 cm, 85°, 28°
9. 4·9 cm, 21°, 65°
10. 55 mm, 62°, 75°

103 Triangle measurement

Remember   The angles inside a shape are called the *interior angles*.

Example   Use a protractor to measure each of the three interior angles inside this triangle. Write these 3 angles down and then add them up.

$\angle BAC = \angle A = 100°$
$\angle ABC = \angle B = 50°$
$\angle BCA = \angle C = 30°$

$\overline{180°}$

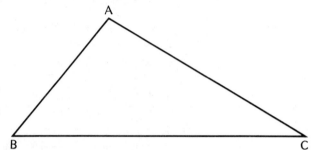

Exercise 1   Use a protractor to measure the interior angles of each of the following triangles, and then find their sum.

1.

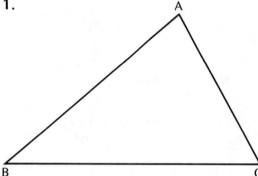

2.

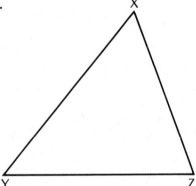

3.

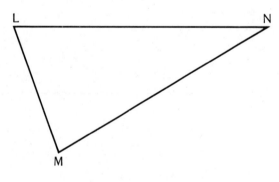

4.

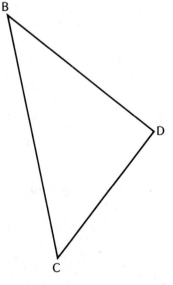

5.

6.

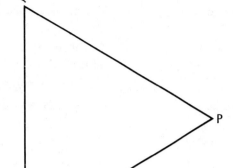

7.

8.

9.

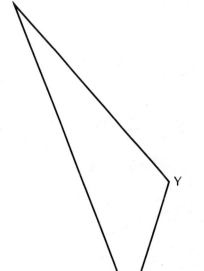

10.

123

104 Angle sum of a triangle

Remember

If you look back at the answers to the exercise in the last section, you should find that the sum of the angles of each triangle was close to 180°
The interior angles of a triangle *always* add up to 180°.
This means that if you know 2 of the angles of a triangle, you can find the third angle by calculation.

Example

Calculate ∠A in the triangle below.

$$\angle B = 60°$$
$$\angle C = 40°$$
$$\overline{\quad\quad}$$
$$100°$$

so, ∠A = 180° − 100°
= **80°**

Exercise 1

Calculate the angle marked with an asterisk * in each triangle and check your answer by measurement:

1.

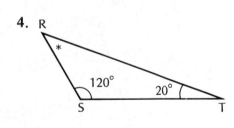

2.

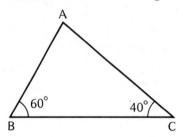

3.

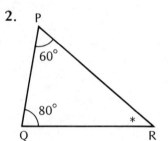

4.

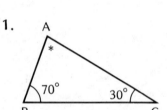

5.

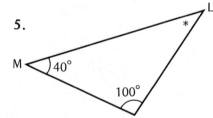

6.

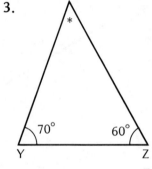

7.

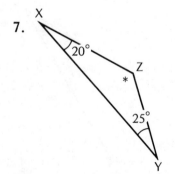

8.

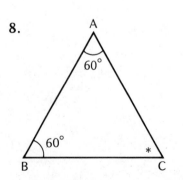

9.

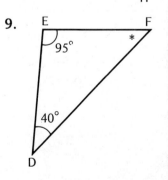

Remember   The same method works even if the triangle contains an obtuse angle:

Examples   1. In a triangle PQR, below, ∠P = 23°, and ∠R = 37°. Find ∠Q.

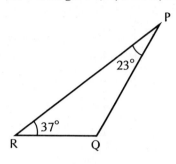

$$\angle P = 23°$$
$$\angle R = 37°$$
$$\overline{\qquad}$$
$$\underline{60°}$$

so, ∠Q = 180° − 60°
= **120°**

2. In a triangle XYZ, ∠X = 54°, and ∠Y = 24°. Find ∠Z.

$$\angle X = 54°$$
$$\angle Y = 24°$$
$$\underline{\qquad}$$
$$\underline{78°}$$

so, ∠Z = 180° − 78°
= **102°**

Exercise 2   Find the missing angle in each of the following triangles:

1.

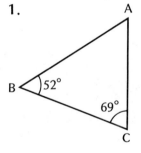

2.

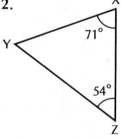

3.

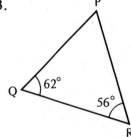

4.

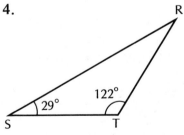

5.

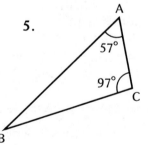

6.

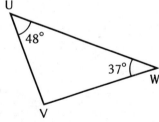

7. In a triangle ABC, ∠A = 64°, and ∠C = 15°. Find ∠B.

8. In a triangle LMN, ∠L = 56°, and ∠N = 24°. Find ∠M.

9. In a triangle XYZ, ∠Y = 63°, and ∠Z = 77°. Find ∠X.

10. In a triangle RST, ∠S = 97°, and ∠T = 31°. Find ∠R.

11. In a triangle JKL, ∠J = 24°, and ∠L = 112°. Find ∠K.

12. In a triangle PQR, ∠P = 116°, and ∠Q = 24°. Find ∠R.

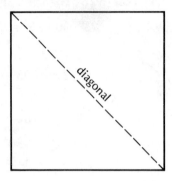 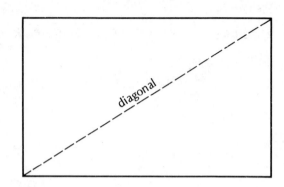

Remember

Looking at the shapes, you can see that:

All the interior angles of squares and rectangles are right angles (90°).

All the sides of the square are the same length.

For the rectangle, opposite pairs of sides are the same length.

A *diagonal* is a line drawn between opposite corners of a rectangle or a square.

Exercise 1

1. Draw a square with side-length 4 cm. You may need a protractor to make sure that you have good right angles.

2. a. Draw a square, side-length 3 cm.
   b. Draw in both diagonals.
   c. Measure each diagonal.
   d. Are the diagonals equal in length?

3. a. Draw a rectangle with side-lengths 3 cm and 4 cm.
   b. Draw in both diagonals.
   c. Measure each diagonal.
   d. Are the diagonals equal in length?

4. a. Draw a square, side-length 5 cm.
   b. Draw in both diagonals.
   c. Measure them.
   d. Are the diagonals equal in length?
   e. What is the angle between the diagonals at the point where they meet?
   f. Do you need to measure this angle with a protractor?

5. a. Draw a rectangle with side-lengths 30 mm and 60 mm.
   b. Draw in both diagonals.
   c. Measure them.
   d. Are the diagonals equal in length?
   e. What is the smaller of the angles between the diagonals at the point where they meet?
   f. What is the larger of the angles between the two disgonals at the point where they meet?
   g. Did you need to measure these two angles with a protractor?

106 Squares and rectangles: perimeter

Remember

The *perimeter* of a shape is the distance around its edge.

Examples

Find the perimeter of the shapes below:

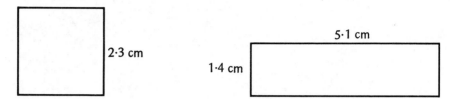

Measure to find that
side-length = 2·3 cm
so, **perimeter** = (2·3 + 2·3 +
               2·3 + 2·3) cm
          = 4 X 2·3 cm
          = **9·2 cm**

Measure to find that
side-lengths = 5·1 cm and 1·4 cm
so, **perimeter** = (2 X 5·1 cm) +
                (2 X 1·4 cm)
          = 10·2 cm + 2·8 cm
          = **13·0 cm**

Exercise 1

Find the perimeter of the following rectangles and squares:

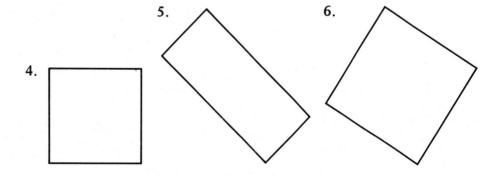

7. A square with side-length 45 mm.

8. A rectangle with side-lengths 34 mm and 70 mm.

9. A rectangle with side-lengths 1·2 cm and 0·8 cm.

10. A rectangle with side-lengths 5 m and 10 m.

107 Squares and rectangles: area

Remember
The *area* of a shape tells you how much surface it has. Areas can be measured in mm², cm², and m²:

1 mm²

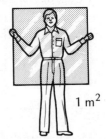

1 cm²

1 m²

The area of a rectangle is found by multiplying together its length and its breadth. The same is true for a square, except that the length and the breadth are the same.

Example
Find the area of the shapes below.

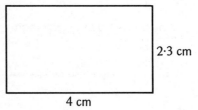

2·3 cm

4 cm

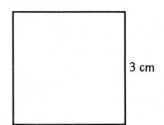

3 cm

area = 4 cm X 2·3 cm
    = **9·2 cm²**

area = 3 cm X 3 cm
    = **9 cm²**

Exercise 1
Find the area of the following rectangles and squares in cm².

1.
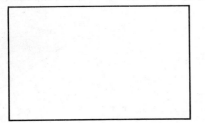

2.
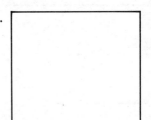

3.

4.

5.
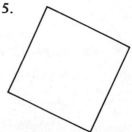

6.
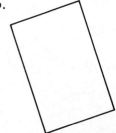

7. Find the area of a square room, whose side length is 9 m.

8. A square roof has 10 equally sized square tiles fitted along one side. How many tiles are needed to cover the roof completely?

Exercise 2

Find the area of the following shapes. They are all made up of rectangles and squares.
Where a shape has a shaded section, that section is not to be included as part of the area.
For each shape you will need to find the area of each rectangle or square separately, and then calculate the actual area.

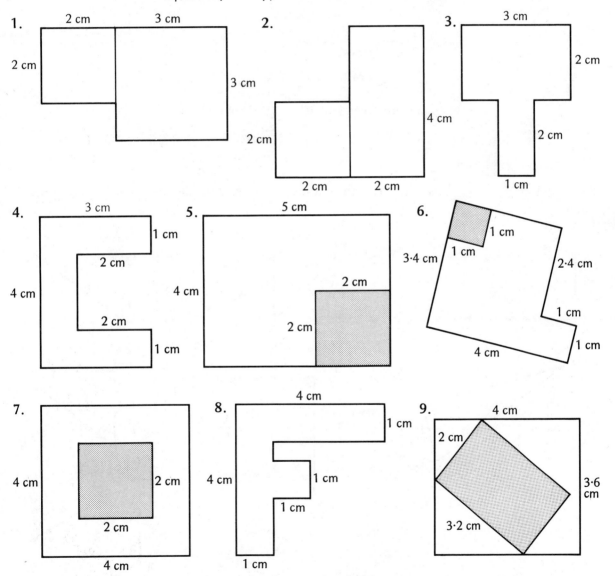

10. You know that 10 mm = 1 cm. Use this square to find how many mm² = 1 cm². You need to find the area of the square in mm² and cm².

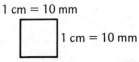

$1 \text{ cm} = 10 \text{ mm}$

$1 \text{ cm} = 10 \text{ mm}$

108 Triangles: area

Remember

A *right-angled triangle* is a triangle with a right-angle at one of its corners.

You can find the area of a right-angled triangle very easily once you have seen that any right-angled triangle is a half of a rectangle.

Example

Find the area of the triangle ABC.

First, copy the triangle and make it a rectangle:

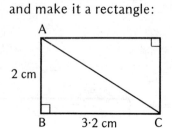

$$\text{Area of rectangle} = 3 \cdot 2 \text{ cm} \times 2 \text{ cm}$$
$$= 6 \cdot 4 \text{ cm}^2$$
so, **area of triangle ABC** $= \frac{1}{2} \times 6 \cdot 4 \text{ cm}^2$
$$= \frac{6 \cdot 4}{2} \text{ cm}^2$$
$$= \mathbf{3 \cdot 2 \text{ cm}^2}$$

Exercise 1

Use a drawing to find the area of each of these right-angled triangles:

1.
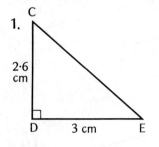

2.
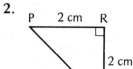

3.
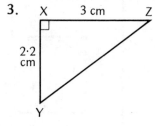

4.
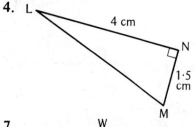

5.
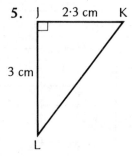

6.
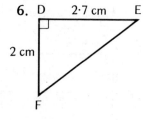

7.
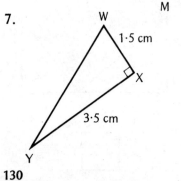

8.
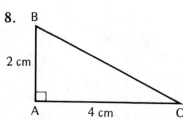

9.
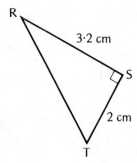

Exercise 2   These sketches are not to scale. Calculate the area of each triangle using

area $= \dfrac{1}{2} \times$ (length of one side) $\times$ (length of other side)

1.
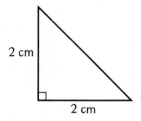
2 cm
2 cm

2.
5 cm
6 cm

3.
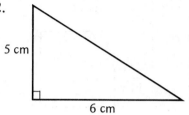
8 cm
12 cm

4.
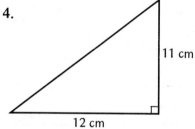
11 cm
12 cm

5.
1·2 cm
6 cm

6.
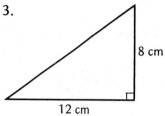
2·5 cm
8 cm

7.
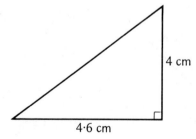
4 cm
4·6 cm

8.
3 cm
5·8 cm

9.
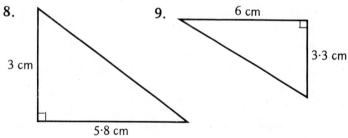
6 cm
3·3 cm

10.
7 cm
4·7 cm

11.
8 cm
2·8 cm

12.
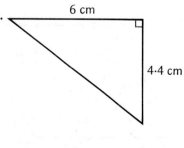
6 cm
4·4 cm

109 Area and perimeter

Exercise 1 Find the perimeter *and* the area of each of the following shapes:

1.

5 cm

2 cm

2 cm

2 cm

2 cm

2.

1·5 cm

1 cm

1·5 cm

1 cm

1 cm

2 cm

3.

3 cm

2 cm

4·5 cm

1 cm

4.

2 m

2 m

2 m

2 m

2 m

5.

2 m

2 m

2 m

2 m

2 m

6.

3 m

1m

1 m

2 m

3 m

1 m

1 m

Exercise 2 Find the area of each of the following shapes:

1.

1 cm

3 cm

5 cm

2.

3 cm

2 cm

3 cm

2 cm

2 cm

3.

3 cm

2 cm   2 cm

4.

2 cm

3 cm

2 cm   2 cm

5.

5 cm

3 cm

3 cm

6.

3 cm

1 cm

2 cm

1 cm   1 cm   1 cm

132

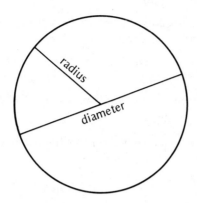

Remember   **diameter** = 2 X **radius** $(d = 2r)$, and **radius** $=\frac{1}{2}$ X **diameter** $(r = \frac{1}{2}d)$
To draw a circle, simply set a pair of compasses to the length of its radius.

Exercise 1   Find, by measuring, the radius of each of these circles, in centimetres.

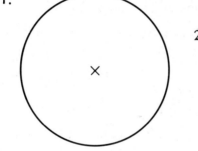

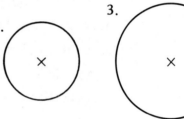

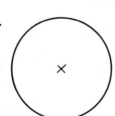

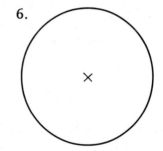

7. Find the diameter of each of the circles above by calculation from the radius that you found. Check your answer by measuring.

Exercise 2   Draw each of the following circles using a pair of compasses:

1a. radius 4 cm   b. radius 3 cm

2a. radius 2·5 cm   b. radius 4·8 cm

3a. diameter 10 cm   b. diameter 8 cm

4a. diameter 9 cm   b. diameter 4·6 cm

111 Circles: circumference

Remember

The distance around a circle is called its *circumference*:
circumference = π X diameter, or C = πd

π is the Greek letter *pi* which in Mathematics stands for:
3·14159265358979323846264338327950288419716939993751 0582...

π *cannot* be written down exactly. For most needs,
3·14 is accurate enough!

Examples

1. Find the circumference of a circle with a diameter of 3 cm.
$C = \pi \times d$
$= 3 \cdot 14 \times 3 \text{ cm}$
= 9·42 cm

2. Find the circumference of a circle which has a radius of 3 m.
$C = \pi \times d$
$= 3 \cdot 14 \times 6 \text{ m}$ (because $d = 2r$)
= 18·84 m

Exercise 1

Find the circumference of each of the following circles:

1a. A circle with diameter 2 mm   **b.** A circle with diameter 5 mm

2a. A circle with diameter 10 mm   **b.** A circle with diameter 6 cm

3a. A circle with diameter 9 cm   **b.** A circle with diameter 12 cm

4a. A circle with radius 2 mm   **b.** A circle with radius 4 mm

5a. A circle with radius 7 cm   **b.** A circle with radius 8 cm

6. Measure the diameter of each of circle below in mm and calculate the circumference.

a.  **b.**  **c.** **d.**

7. Find the distance round a circular pond which has a radius of 5 m.

8. Find the circumference of a car wheel which has a diameter of 0·5 m.

9. A can of baked beans has a diameter of 10 cm. Find the circumference of the tin.

10. Can you work out a way to check your answers to question 6 by measuring?

112 Circles: area

Remember

To find the area of a circle you need to know the following formula:
area of circle = π × (radius)²
or, $A = \pi r^2$

Examples

1. Find the area of a circle which has a radius of 2 cm.
$A = \pi r^2$
$= 3 \cdot 14 \times 2 \times 2$
$= 12 \cdot 56 \text{ cm}^2$

2. Find the area of a circle with a diameter of 6 mm.
$A = \pi r^2$
$= 3 \cdot 14 \times 3 \times 3$ (if $d = 6$, then $r = 3$)
$= 28 \cdot 26 \text{ mm}^2$

Exercise 1

Find the area of the following circles:

1a. A circle, radius 1 mm   b. A circle, radius 4 cm

2a. A circle, radius 5 cm   b. A circle, radius 6 mm

3a. A circle, radius 8 mm   b. A circle, radius 9 cm

4a. A circle, diameter 4 m   b. A circle, diameter 6 cm

5a. A circle, diameter 14 cm   b. A circle, diameter 20 mm

6. Measure the diameters of the following circles, then find their areas:

a.    b.    c.    d.

7. Find the area of a circular pond with a diameter of 30 m.

8. Find the area of a motor cycle wheel which has a radius of 28 cm.

9. Find the area of a tractor wheel which has a diameter of 120 cm.

10. Find the area of a dust bin lid which has a radius of 72 cm.

11. Find the area of a foreign coin which has a diameter of 2 cm.

12. Find the area of a ginger snap which has a radius of 3 cm.

13. Find the area of a discus which has a diameter of 19 cm.

14. Find the area of a custard pie which has a diameter of 16 cm.

113 Maps: scale and distance

Here is a map showing the position of the home grounds of some football teams.

The *scale* of the map is
1 cm to 100 km

Example

How far must Southampton FC travel in order to play a match against Newcastle?

Distance on the map = 4·5 cm

so, actual distance = 4·5 × 100 km

= **450 km**

Exercise 1

In the following football matches, find how far the away team has had to travel to the nearest 10 km:

1.	Newcastle United	v	Nottingham Forest
2.	Oxford United	v	Cardiff City
3.	Glasgow Rangers	v	Aberdeen
4.	Norwich City	v	Liverpool
5.	Leeds United	v	Plymouth
6.	Middlesbrough	v	Manchester City
7.	Manchester United	v	Brighton
8.	Aberdeen	v	Plymouth
9.	Cardiff	v	Newcastle United
10.	Liverpool	v	Preston

All questions on this page refer to the map in the previous section.

Remember

These are the major points of the compass. You must *remember* where **North, South, East and West** are. All the others you can work out.

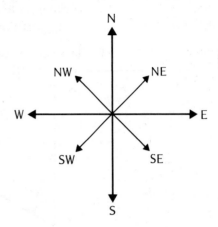

Examples

1. Which town lies (approximately) due South of Nottingham?
   Oxford

2. Which club is about 500 km NW of Norwich?
   Glasgow Rangers

Exercise 1

1. Which town lies (approximately) due East of Southampton?

2. Which town lies due South of Middlesbrough?

3. Which football club's home ground is (approximately) due West of Brighton?

4. Which football club's home ground is due East of Liverpool?

5. Which town is roughly 210 km NE of Cardiff?

6. Which town is roughly 140 km NW of Nottingham?

7. Which town is roughly 130 km SE of Oxford?

8. Which town is roughly 160 km SW of Middlesbrough?

9. If Brighton FC set off in a NW direction and travel for 130 km, find out where they are playing.

10. Nottingham Forest set off in the SW direction and travel for 220 km. Who are they playing?

11. Brighton travel 90 km to a match. Who are they playing?

12. When Manchester City travel due West, who are they playing?

13. If Aberdeen travel 200 km SW, where are they going?

14. How far will Plymouth need to travel to play against Norwich?

15. How far do Oxford travel when they play a game against Leeds?

115 Bearings

Remember

A *bearing* is an accurate way of giving the direction of one place from another. A bearing is an angle — it is the angle in a clockwise direction from North..Bearings are always written as 3-figure numbers:

Examples

1. The bearing of A from O is 060°   2. The bearing of B from O is 135°

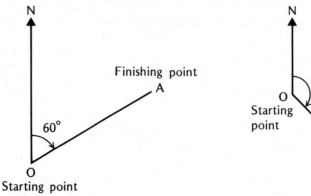

Exercise 1

In each question, find the bearing of C from O.

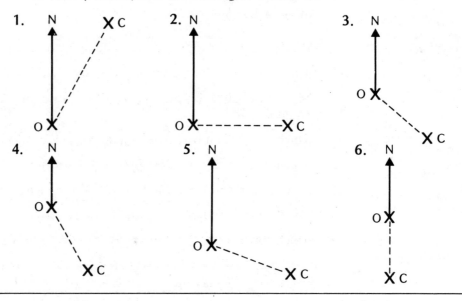

Example 1

Find the bearing of D from O.

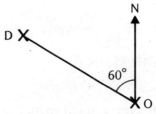

The angle between North and D is 60°, but in the wrong direction. The angle that we need is 360° − 60° = 300°, because a whole turn = 360°. So, **the bearing of D from O = 300°.**

Example 2   Find the bearing of E from O.

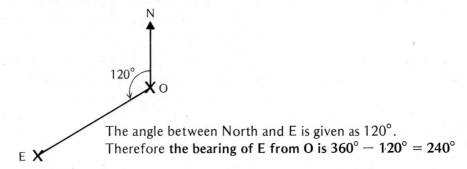

The angle between North and E is given as 120°.
Therefore **the bearing of E from O is 360° − 120° = 240°**

Exercise 2   In each question, find the bearing of F from O.

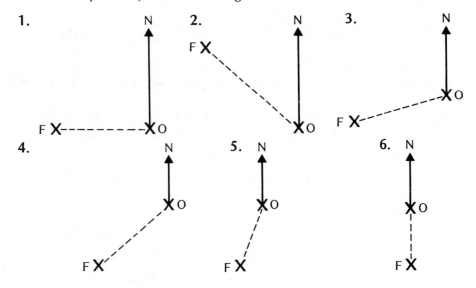

Exercise 3   Use the map in section 113 to answer the following questions.
Your answer only needs to be accurate to the nearest 5°:

1. Find the bearing of Norwich *from* Oxford.

2. Find the bearing of Middlesbrough *from* Liverpool.

3. Find the bearing of Aberdeen *from* Glasgow.

4. Find the bearing of Newcastle *from* Glasgow.

5. Find the bearing of Norwich *from* Leeds.

6. Find the bearing of Southampton *from* Brighton.

7. Find the bearing of Oxford *from* Norwich.

8. Find the bearing of Liverpool *from* Brighton.

9. Find the bearing of Glasgow *from* Newcastle.

10. Find the bearing of Nottingham *from* Southampton.

116   Bearings: scale drawings

Remember

When given a route that involves bearings it is best to make a *scale drawing*. This is where the drawing is the shape, but many times smaller than the real thing.

Example

Spike, the orienteer, runs 2 km on a bearing of 060°, and then changes direction onto a bearing of 120° and runs another 3 km. How far is he from his starting point? Use a scale of 1 cm to 1 km.

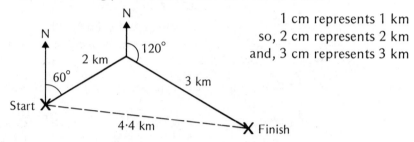

1 cm represents 1 km
so, 2 cm represents 2 km
and, 3 cm represents 3 km

Measure: start to finish = 4·4 cm (roughly)
so, **true distance from starting point = 4·4 km (roughly)**

Exercise 1

Make scale drawings of each of the following journeys. Find the distance of the finishing point from the starting point in each case. You are given the scales.

1. A plane flies 300 km on a bearing of 070° and then changes direction onto a bearing of 110°, flying another 500 km. Use a scale of 1 cm to 100 km.

2. A ship travels 400 km on a bearing of 110°, then 600 km on a bearing of 060°. (Scale: 1 cm to 100 km).

3. A herd of buffalo travel 3 kilometres on a bearing of 080°, then 4 kilometres on a bearing of 170°. (Scale: 1 cm to 1 km).

4. A balloon floats 64 km on a bearing of 130° then, after a wind change, another 36 km on a bearing of 210°. (Scale: 1 mm to 1 km).

5. A helicopter flies 36 km on a bearing of 200° and then 18 km on a bearing of 300°. (Scale: 1 cm to 6 km).

6. A hovercraft travels 60 km on a bearing of 075° and then turns to a bearing of 110° and travels 42 km. (Scale: 1 mm to 1 km).

7. A bird flies 28 km on a bearing of 135° then 18 km on a bearing of 020°. (Scale: 1 mm to 1 km).

8. A motor cruiser travels 360 km on a bearing of 026° then 400 km on a bearing of 300°. (Scale: 1 mm to 10 km).

117 Cubes and cuboids

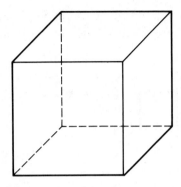 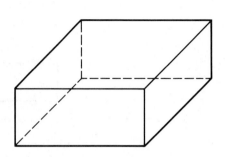

Remember

A cube has six sides.
All sides are exact squares.
All corner angles are 90°.

A cuboid has six sides.
All sides are rectangles,
which can be squares.
All corner angles are 90°.

Exercise 1

The following everyday objects are cubes or cuboids.

1. Make a list of the cubes.

2. Make a list of the cuboids that are not cubes.

   A. shoe box
   B. text book
   C. building bricks
   D. tooth paste box
   E. tea packet
   F. a box containing a football
   G. dominoes
   H. dice

   I. match book
   J. sugar lumps
   K. a box 4 cm by 4 cm by 3 cm
   L. a box 3 cm by 3 cm by 3 cm
   M. 'OXO' cube
   N. a pack of butter
   O. Rubik's cube

Exercise 2

1. Write down 3 things that are cuboids that are not listed above.

2. Write down 3 things that are cubes that are not listed above.

3. How many edges has a cube?

4. How many edges has a cuboid?

5. How many corners has a cube?

6. How many corners has a cuboid?

7. How many sides of a solid cube can you see at one time?

8. How many sides of a solid cuboid can you see at one time?

118 Cubes and nets

Remember

The diagram below is a *net of a cube.* If you were to cut it out of paper or card and fold it into a solid shape, that would be a cube.

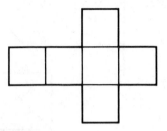

Exercise 1

Which of the following are nets of a cube? You should draw each one, cut it out and see if it makes a cube.

1. 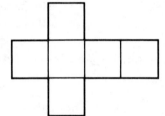   2.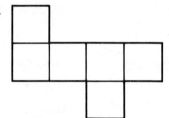

3. 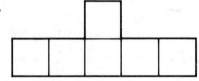   4.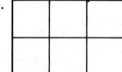

5. 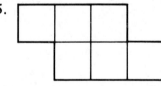   6.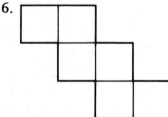

7. Make up three nets for a cube.

8. Make up three nets for a cuboid.

9. Draw a net for a cube with side-length 15 mm.

10. Draw a net for a cuboid with side-lengths 20 mm, 30 mm and 40 mm.

Remember

Distance measures length   [units mm, cm, m]
Area measures surface   [units mm^2, cm^2, m^2]
Volume measures space taken up   [units mm^3, cm^3, m^3]

1 mm^3

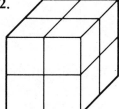

volume: 1 cm^3

1 cm   1 cm   1 cm

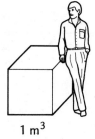

1 m^3

Exercise 1

Work out how many cm^3 blocks there are in each solid shape. You must remember to count any blocks that are hidden!

1.
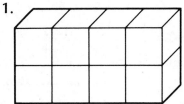

2.
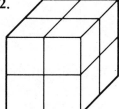

3.
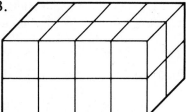

4.
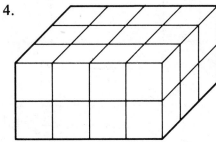

5.
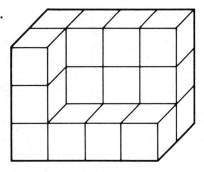

6.
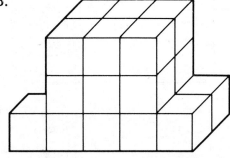

7.
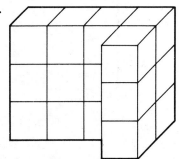

120 Volume calculations

Remember

You can find the volume of a simple shape by calculation:

Examples

1. Find the number of blocks
in this group by calculation.
Find the volume if each
block is a centimetre cube.

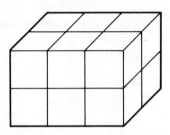

Width = 3 blocks, length = 2 blocks, height = 2 blocks
so, number of blocks $\quad= 3 \times 2 \times 2$
$\qquad\qquad\qquad\qquad= 12$
so, **volume of group** $\quad= \mathbf{12\ cm^3}$

2. Find the volume of a group of blocks, width 3 cm, length 3 cm and
height 1 cm
volume $= 3 \times 3 \times 1$
$\qquad\quad= \mathbf{9\ cm^3}$

Exercise 1

Find the volume of each of the following cuboids:

1.

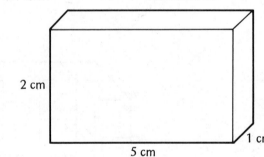

2 cm

5 cm

1 cm

2.

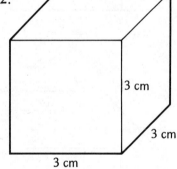

3 cm

3 cm

3 cm

3 cm

3.

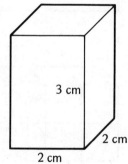

3 cm

2 cm

2 cm

4.

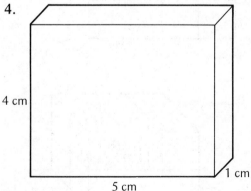

4 cm

5 cm

1 cm

Exercise 2

Find the volume of each of the following cuboids. You must be careful with the units — a volume is always a unit *cubed*, like cm³, m³, or mm³.

1.

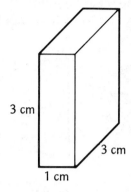

3 cm

3 cm

1 cm

2.

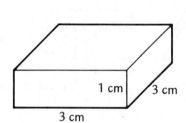

1 cm 3 cm

3 cm

3.

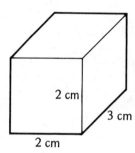

2 cm

3 cm

2 cm

4.

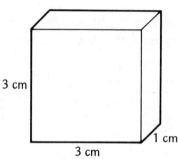

3 cm

1 cm

3 cm

5.

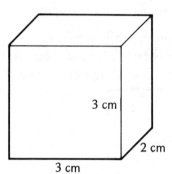

3 cm

2 cm

3 cm

6.

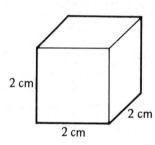

2 cm

2 cm

2 cm

7.

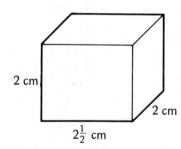

2 cm

2 cm

$2\frac{1}{2}$ cm

8.

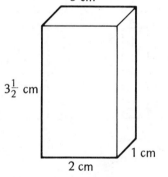

$3\frac{1}{2}$ cm

1 cm

2 cm

9.

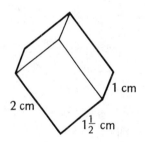

2 cm

1 cm

$1\frac{1}{2}$ cm

Exercise 3

Copy and complete this table of volumes of cuboids. The first one has been done for you as an example.

	length	width	height	volume
1.	6 cm	5 cm	2 cm	60 cm³
2.	10 mm	2 mm	2 mm	
3.	2 m	10 m	2 m	
4.	6 km	3 km	1 km	
5.	15 cm	2 cm	3 cm	
6.	3 cm	9 cm	1 cm	

121 Line symmetry

Remember

If a shape can be folded so that one part covers the other part *exactly*, then the fold is on a *line of symmetry*. Some shapes have no lines of symmetry, others have just one line of symmetry and some have more than one line of symmetry.

Examples

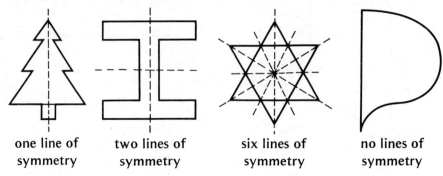

| one line of symmetry | two lines of symmetry | six lines of symmetry | no lines of symmetry |

Exercise 1

Copy each of the following shapes and draw in the lines of symmetry. Write underneath how many lines of symmetry the shape has. Some have no lines of symmetry at all.

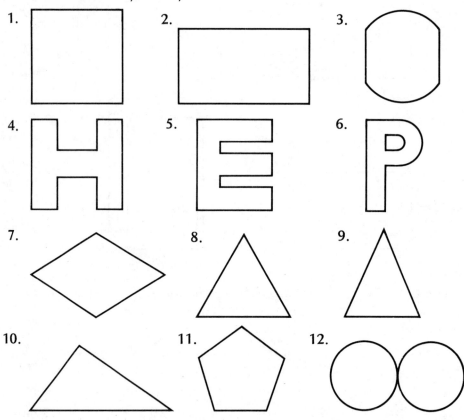

Exercise 2

1. Draw a shape of your own that has 1 line of symmetry.

2. Draw a shape of your own that has 2 lines of symmetry.

3. Draw a shape of your own that has no lines of symmetry.

122 Tessellations

Remember A tessellation is a way of covering a surface with similar shapes that fit together and leave no gaps between them:

Examples

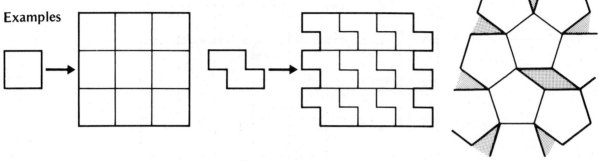

squares *do* tessellate:
they leave no gaps

this shape tessellates:
it leaves no gaps

pentagons *do not* tessellate: they leave **gaps**

Exercise 1 Do repeat drawings of each of the following shapes to see if they tessellate. If you have cardboard cut-out shapes, these will help.

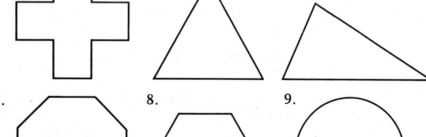

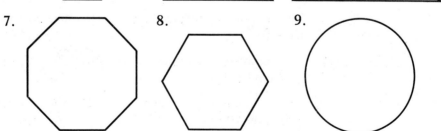

10. Draw a shape of your own that tessellates.

11. Draw a shape of your own that does *not* tessellate.

12. Do all the shapes above that *do* tessellate have at least one line of symmetry?

This crossword is based on work done on geometry. Copy the crossword and fill in the answers to the clues.

ACROSS

1. Find the area of a 6 cm square.

2. The perimeter of a rectangle, 6 m by 7 m.

4. The area of a right-angled triangle base 10 m and height 5 m.

6. The volume of a cube, side 10 mm.

8. The area of a rectangle, 10 m by 8 m.

10. The volume of a cuboid, 23 cm by 4 cm by 2 cm.

12. The perimeter of a rectangle, 200 m by 102 m.

14. The area of a rectangle, 30 cm by 7 cm.

15. The area of a rectangle 100 m by 6 m.

DOWN

1. The area of a rectangle, 12 cm by 3 cm.

2. One right angle plus 118°.

3. The perimeter of a field, 180 m by 120 m.

5. 40° less than a right-angle.

7. 11° more than 3 right-angles.

9. The volume of a cuboid, 1 mm by 40 mm by 20 mm.

10. The area of a right-angled triangle, base 12 mm, height 2 mm.

11. The perimeter of a rectangle, 16·5 mm by 3·5 mm.

12. The area of a rectangle, 12 cm by $5\frac{1}{2}$ cm.

13. The perimeter of a square, side 10 mm.

ACROSS

16. The area of a right-angled triangle with base 7 cm and height 6 cm.

19. The number of degrees in a half-turn.

21. The number of degrees in a right-angle.

22. The perimeter of a square side 16 cm.

23. The area of a rectangle 12 mm by 6 mm.

DOWN

16. The perimeter of a rectangle, 63 m by 45 m.

17. 4° more than the angle in a half-turn.

18. The perimeter of a triangle with sides, 8 cm, 9 cm and 22 cm.

20. The perimeter of an 8 m square.

Geometry: **Revision test**

1. Using a ruler, draw straight lines, then mark off the following lengths:
   a. 16 mm   b. 36 mm   c. 2·4 cm   d. 4·6 cm   e. 5·1 cm

2. Using ruler and compasses construct a triangle with sides of:
   a. 26 mm, 34 mm and 41 mm
   b. 3·6 cm, 2·1 cm and 1·9 cm
   c. 2·7 cm, 42 mm and 3·4 mm

3. Work out the angle, in degrees, turned in each case:
   a. A half turn followed by a quarter turn.
   b. A quarter turn followed by a quarter turn.
   c. A half turn, followed by a quarter turn, followed by a half turn in the opposite direction.

4. Use your protractor to draw the following angles:
   a. 70°   b. 83°   c. 124°   d. 220°   e. 270°

5. In triangle ABC, angle A = 37°, angle B = 65°. Find the size of angle C.

6. a. Draw a square of side 2·8 cm.
   b. Draw in both diagonals.
   c. Measure them and write down their lengths.
   d. Are they equal? Should they be?

7. Find the area and perimeter of a rectangle which is 4·3 cm long and 5 cm wide.

8. Construct the triangle shown exactly into your book using ruler, compasses and protractor. When you have finished, write in the size of A.

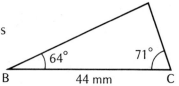

9. Construct triangle ABC such that AB = 3·6 cm, AC = 41 mm and angle A = 67°. When you have finished write down the size of BC.

10. Find the areas of the two triangles shown:

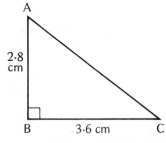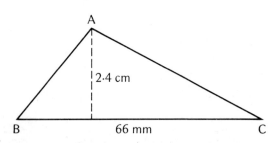

11. A plane flies 400 km on a bearing of 075°, then turns on to a bearing of 165° and flies 280 km. By constructing a scale drawing, find how far it is from its starting point. (Scale 1 cm to 100 km.)

12. A bicycle wheel has a diameter of 20 cm. What is its area? [π = 3·14]

13.

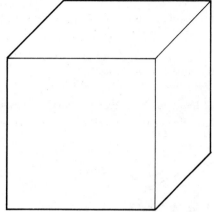

    a. The front of the cube shown above is drawn exactly. Measure the length and then find the area of the front.

    b. How many faces does the cube have? Use this and your answer to part **a**) to find the surface area of the cube.

    c. Draw a net for this cube.

    d. Find the total area of the net.

    e. What do you notice about the two areas in **c** and **d**?

14. Find the number of blocks in the following shapes:

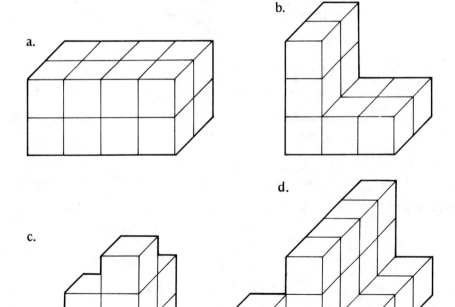

a.

b.

c.

d.

15. Find the volume of a rectangular trunk which has a length of 120 cm, a width of 100 cm and a height of 65 cm.

Remember

Algebra can be used to simplify things. These two questions are the same.
1. John has 4 pairs of jeans and 8 shirts in the wash. He has 2 pairs of jeans and 3 shirts in his wardrobe. He is wearing one of each. How many jeans and shirts altogether?
2. Simplify $4j + 8s + 2j + 3s + j + s$

The answer to the first sum is 7 pairs of jeans and 12 shirts.
The answer to the second is $7j + 12s$

Examples

1. Simplify $4w + 6t + 2w + 2t$
$4w + 6t + 2w + 2t$
is the same as $6w + 8t$

2. Simplify $3d + 5d + 6c + c$
(c written by itself means $1c$)
$3d + 5d + 6c + c$
is the same as $8d + 7c$

Exercise 1

Simplify each of the following:

1a. $4a + 3b + 5a + 8b$ b. $6j + 10k + 11j + 6k$

2a. $12c + 11d + 8c + 4d$ b. $4d + 6e + 14d + 8e$

3a. $2m + n + m + 4n$ b. $17h + 12j + 2h + j$

4a. $3v + w + 5v + 2w$ b. $7r + 5s + 8r + s$

5a. $4r + 6r + 2w + w$ b. $7a + 8a + 3t + t$

6a. $u + 3u + 10v + 3v$ b. $4p + p + 2q + q$

7a. $h + 5h + 4j$ b. $9d + 3e + 3e + 2d$

8a. $6t + 9t + w$ b. $s + 4t + 2t$

9a. $5c + 4d + d + c$ b. $9f + g + g + f$

10a. $8a + 5g + 3g + a + a$ b. $s + t + t + 2t + 9s$

Remember

It makes no difference if there are more than two letters — just group together each individual letter.

Exercise 2

1. $7a + 8b + 10c + 3a + 6b + 8c$

2. $13x + 14y + 6z + 12z + 15y + 11x$

3. $24p + 18q + 12r + 6q + 2p + r$

4. $2b + c + 4d + c + d + 9b + c + 3c$

5. $5r + t + 5s + 3r + 5r + 8s + 8t$

6. $4r + 6t + 3s + 3r + 6t + 8s$

7. $3d + 4e + 8f + 2f + 4d + 9e$

Remember

Simplification can involve subtraction as well as addition. These two questions are the same.
1. Out of a collection of 35 LPs and 40 singles, John lent the school disco 12 LPs and 10 singles. How many did they have left at home?
2. Simplify $35L + 40S - 12L - 10S$
The answer to the first sum is 23 LPs and 30 singles.
The answer to the second is $23L + 30S$.

Examples

1. Simplify $4a + 5t - 2a - 3t$
$4a + 5t - 2a - 3t$
is the same as $2a + 2t$

2. Simplify $7m + 2m - m + 3t - 2t$
$7m + 2m - m + 3t - 2t$
is the same as $8m + t$

Exercise 3

Simplify each of the following:

1a. $3a + 2b - 2a - b$ b. $2c + 6d - c - 4d$

2a. $9p + 3q - 3p - q$ b. $5g + 3h - 2g - 3h$

3a. $10x + 10y - 5x - 5y$ b. $6d + 4e - 4d - 2e$

4a. $16a + 14b + 10a - 8b$ b. $4d + 18e + 23d - 6e$

5a. $14k + 7m + 3n - 2k - 4m$ b. $8a + 6b + 7c + 2a - 4c + 3b$

6a. $5p + 4p - 3p$ b. $3q + q - 2q$

7a. $12r + 8s + 7s - 5r$ b. $23m + 16n - 8n + 7m$

8a. $24a + 16b - 18a - b$ b. $8x + 4y + 4x - 2y$

9a. $5a + 7b - 3a + 6c - 3c$ b. $22c + 40d + 20d - 20c$

10a. $5a + 12b + 28c - 3a - 22c - 5b$ b. $22x + 19y + 16z - 11y + 8z - 10x$

Example

Simplify $6a + 3b - 2b - 10b$
$6a + 3b - 2b - 10b$
is the same as $6a - 9b$

Exercise 4

Simplify each of the following:

1a. $10a + 3b + 3a - 6b$ b. $12c + 6d - 3c - 8d$

2a. $6b + 6d - 2b - 10d$ b. $14l + 10m + l - 16m$

3a. $12c + 4e - 6c - 6e$ b. $8m + 14n - 5m - 20n$

4a. $14d + 2p - 7d - 5p$ b. $6p + 7q + p - 10q$

5a. $6e + 8f + 3e - 10f$ b. $11s + 10t - 7s - 12t$

6a. $5x + 7y + 2x - 8y$ b. $10t + 12s - 2t - 14s$

7a. $3b + 3d - d - 4d$ b. $6d + 16f - 3d - 17f$

8a. $8l + 8m - l - 10m$ b. $11e + 12g + 3e - 13g$

Remember 2 × n can be written as $2n$ and n × 3 can be written as $3n$. Notice how the number is always written at the front.

Examples
1. Use algebra to write in a shorter way:
   a. a multiplied by 3
       a × 3 or $3a$
   b. 2 multiplied by x
       2 × x or $2x$

2. Use algebra to write in a shorter way:
   a. c minus 4
       $c - 4$
   b. 5 added to d
       $d + 5$

Exercise 1 Use algebra to write in a shorter way:

1. d multiplied by 5   2. 2 multiplied by z

3. Twice v 4. 10 multiplied by x

5. 6 added to a 6. 4 multiplied by b

7. p minus 8   8. z plus four

9. a multiplied by z 10. c added to a

Remember n × n or nn can be written as n^2. n^2 is read "n squared".
p × p × p can be written as p^3. p^3 is read "p cubed".

Examples
1. p × p × q × q × q
   is the same as p^2 × q^3
   is the same as $p^2 q^3$

2. 3 × s × 3 × s × s
   is the same as 3 × 3 × s × s × s
   is the same as $9s^3$

Exercise 2 Rewrite each of the following in a shorter way:

1a. q × q b. z × z c. r × r × r d. 3 × 3

2a. y × y b. k × k c. 5 × 5   d. (ab) × (ab)

3a. $(p$ × $p)$ × q b. 4 × $(a$ × $a)$ c. $(3$ × $3)$ × $(c$ × $c)$ d. p × p × p

4a. c × c × d × d b. x × y × x × y

5a. m × n × m × n × m b. p × p × p × q × q

6a. 4 × 4 × s × s × s b. 5 × 5 × z × z × z

7a. f × g × g × g × f b. r × s × s × r × s × r

8a. u × v × v × u × u × v b. 3 × a × a × a × 3

9a. 2 × p × 2 × p × 2 × p b. 4 × x × x × 4 × x

10a. 6 × m × m × m × 6   b. 3 × y × 3 × y × y

11a. 10 × t × 10 × t × t b. k × 7 × k × 7 × k

12a. 8 × d × 8 × d × d b. j × j × 9 × j × 9

126 Algebra: order of operations

Remember
In order to simplify an algebraic sentence with more than one operation ($+$, $-$ and $\times$, say) you must simplify the multiply part first. The rest of the simplification is exactly the same as before.

Examples

1. $4a \times 5 + 2a \times 3 + 5a \times 8$
   is the same as $20a + 6a + 40a$
   is the same as **$66a$**

2. $3 \times 4 \times 2w + 3 \times 4w - 3 \times 10w$
   is the same as $24w + 12w - 30w$
   is the same as **$6w$**

Exercise 1
Simplify the following examples:

1a. $6a \times 7 + 8a$ b. $5a \times 6 + 16a$

2a. $6a \times 5 + 4a$ b. $8b \times 6 + 10b$

3a. $8b \times 4 + 6b$ b. $2p \times 3 + 5p$

4a. $2 \times 7s - 8s$ b. $3 \times 5 \times 3c$

5a. $8d + 4d \times 2$ b. $10b + 6b \times 7$

6a. $6s \times 6 - 12s$ b. $30p - 4p \times 4$

7a. $4a \times 5 + 3a \times 7$ b. $6g \times 8 + 8g \times 9$

8a. $6a \times 8 - 3a \times 5$ b. $8c \times 9 - 5 \times 3c$

9a. $12 \times 7m - 4 \times 9m$ b. $13 \times 6c - 25c \times 3$

10a. $15k \times 4 - 7 \times 3k$ b. $45n \times 2 - 2n \times 45$

11a. $5a \times 3 + 6a \times 4 + 8a \times 5$ b. $4 \times 5d + 6 \times 3d + 6 \times 2d$

12a. $3p \times 5 + 5p + 8 - 4 \times 6p$ b. $6b \times 4 + 5b \times 8 - 3b \times 7$

13a. $7b \times 7 + 4 \times 7b - 5b \times 8$ b. $5z \times 8 - 4z \times 5 + 6z \times 7$

14a. $12a \times 8 - 3a \times 5 - 2a \times 3$ b. $5b \times 4 - 7b \times 5 + 6b \times 8$

15a. $12c \times 4 - 8 \times 9c + 10 \times 4c$ b. $20f \times 5 - 5 \times 9f - 4 \times 4f$

16a. $16d \times 5 - 20d \times 0$ b. $3a \times 4 \times 3 - 2a \times 2 \times 2$

17a. $6d \times 3 \times 6 + 3 \times 4 \times 2d$ b. $7c \times 2 \times 3 - 3 \times 2 \times 3c$

18a. $3 \times 4 \times 2f + 5f \times 3 \times 4$
     b. $2a \times 3 \times 4 + 3 \times 3a \times 5 + 2a \times 5 \times 3$

19a. $4 \times 4c \times 2 + 4 \times 2 \times 3c - 6c \times 2 \times 2$
     b. $7g \times 7 \times 2 - 2 \times 3 \times 2g - 4g \times 2 \times 1$

20a. $3 \times 4 \times 2s - 5s \times 3 \times 4 + 6s \times 3 \times 2$
     b. $4 \times 5b \times 2 - 5b \times 5 \times 4 + 2 \times 2 \times 30b$

127 Algebra: substitution

Example

Substitute $a = 4$ to calculate $8a - 12 + 4a$

$8a - 12 + 4a$

is the same as $(8 \times 4) - 12 + (4 \times 4)$

is the same as $32 - 12 + 16 = 36$

Exercise 1

Substitute $a = 4$ in each of the following expressions:

1a. $2a + 2$ b. $4a + 5$ c. $12 + a$

2a. $2 + 7a$ b. $7a - 9$ c. $6a - 13$

3a. $9a - 22$ b. $3a + 8 + 4a$ c. $7a + 8$

4a. $8a + 1 + 4a$ b. $1 + 12a$ c. $4a + 7a - 18$

If $p = 6$ and $q = 8$, find the values of:

5a. $p + q$ b. $p + 2q$ c. p^2

6a. $2p + 3q$ b. $5p - 3q$ c. $12p - 7q$

7a. pq b. $3pq$ c. $3p^2$

8a. $2p^2 + q^2$ b. $p^2 + 2q^2$ c. $3p + 4q^2$

Example

Substitute $a = 2$ to calculate $2a^2 + a + 1$

$2a^2 + a + 1$

is the same as $(2 \times a \times a) + a + 1$

is the same as $(2 \times 2 \times 2) + 2 + 1 = 8 + 2 + 1 = 11$

Exercise 2

Find the values of:

1. $a^2 + a + 4$ when $a = 2$ 2. $b^2 + 5b + 2$ when $b = 4$

3. $2g^2 + 3g + 3$ when $g = 6$ 4. $m^2 + 6m + 10$ when $m = 6$

5. $p^2 + 7p + 6$ when $p = 4$ 6. $f^2 + 3f + 4$ when $f = 5$

7. $3h^2 + 2h - 2$ when $h = 4$ 8. $4z^2 + z - 4$ when $z = 1$

9. $k^2 + 4 - 4k$ when $k = 2$ 10. $2m^2 + 3m - 3$ when $m = 1$

Examples

1. Substitute $l = 3$ cm in $P = 4l$ to find the perimeter of a square.
$l = 3$ cm $P = 4l$ so
$P = 4 \times 3$ cm
$= 12$ cm

2. Substitute $r = 2$ cm in $A = \pi r^2$ to find the area of a circle.
Take π to be $3 \cdot 14$
$r = 2$ cm $A = \pi r^2$
so $A = \pi \times (2 \text{ cm})^2$
$= 3 \cdot 14 \times (2 \text{ cm} \times 2 \text{ cm})$
$= 3 \cdot 14 \times 4 \text{ cm}^2$
$= 12 \cdot 56 \text{ cm}^2$

3. Substitute $l = 2$ m and $b = 4$ m in $A = lb$ to find the area of a rectangle.
$l = 2$ m and $b = 4$ m
$A = lb$
so $A = 2$ m $\times 4$ m
$= 8 \text{ m}^2$

Exercise 3

1. Substitute $l = 2$ km in $P = 4l$ to find the perimeter of a square field.

2. Substitute $r = 1$ mm in $A = \pi r^2$ to find the area of a circle.

3. Substitute $r = 4$ cm in $d = 2r$ to find the diameter of a circle.

4. Substitute $S = 30$ km per hour and $D = 2$ hours in $D = ST$ to find the distance travelled by a bus.

5. Substitute $l = 20$ mm and $b = 3$ mm in $P = 2l + 2b$ to find the perimeter of a rectangle.

6. In the equation $A = l^2$ substitute $l = 5$ m to find the area of a square.

Example

Find the value of $\dfrac{(b + c)^2 + a}{a}$ if $a = 3$, $b = 1$ and $c = 2$.

Brackets first: $(b + c)^2$ is the same as $(1 + 2)^2$ is the same as 3^2, i.e. **9**

so $\dfrac{(b + c)^2 + a}{9}$ is the same as $\dfrac{9 + 3}{3}$ is the same as $\dfrac{12}{3}$, i.e. **4**

Exercise 4

If $a = 6$, $b = 4$, $c = 8$ and $d = 2$ find the value of each of the following expressions:

1a. $a + c + d$ **b.** $d + 3c + a$

2a. $3a + 3c + 6d$ **b.** $2a + c + d$

3a. $6b + 2c + a$ **b.** $ab + ba$

4a. $ab + bc + 2$ **b.** $cd + 1 + da$

5a. $(ab) \div c$ **b.** $c \div bd$

6a. $3c \div 2a$ **b.** $(a + b + c) \div d$

7a. $\dfrac{a + b + c}{a}$ **b.** $(bc + bd) \div c$

8a. $b^2 + c^2 - a^2$ **b.** $ab + abd$

9a. $\dfrac{(a^2 + c^2)}{b}$ **b.** $\dfrac{c^2 - b}{a}$

10a. $\dfrac{ac + b^2}{c}$ **b.** $2b^2 + 2a^2$

11. An old mariner's rule says that to see n miles to sea, you must have your eyes $\dfrac{2n^2}{3}$ feet above sea level. How many feet must your eyes be above sea level to be able to see   **a.** 3 miles?   **b.** 9 miles?

Remember

Algebra is a shorthand. Each of these questions is very similar:
1. John thinks of a number and adds 2. The answer is 3.
   What was the first number? **1**
2. $x + 2 = 3$. What is the value of x? $x = 1$

The sums were the same but in the second one we have used algebra to solve an equation.

Exercise 1

Write all of these as algebraic equations, then solve them, to find the 'missing number':

1. Susan thinks of a number and adds 6. The total is nine.

2. Yvonne thinks of a number and adds 8. The total is twelve.

3. Mark thinks of a number then adds 10. The total is 14.

4. Andrew thinks of a number, then adds 7. The total is twenty.

5. Veronica thinks of a number, then subtracts 8. The answer is six.

6. Terry thinks of a number, then subtracts eight. The answer is two.

7. Marie thinks of a number, then subtracts 12. The answer is twelve.

8. Tony thinks of a number, then subtracts five. The answer is fifteen.

9. Mary thinks of a number, then subtracts sixteen. The answer is eleven.

10. Tom thinks of a number, then subtracts thirteen. The answer is fifteen.

11. Dick thinks of a number, then adds thirty. The answer is fifty two.

12. Harry thinks of a number, then subtracts twenty three. The answer is nineteen.

Exercise 2

Solve the following equations:

1a. $a + 4 = 5$ b. $z + 6 = 8$

2a. $c + 5 = 9$ b. $x + 7 = 14$

3a. $p + 12 = 19$ b. $q + 15 = 23$

4a. $y + 13 = 25$ b. $m + 11 = 18$

5a. $k + 14 = 24$ b. $r + 23 = 30$

6a. $h + 24 = 39$ b. $k + 24 = 40$

7a. $a - 5 = 11$ b. $x - 4 = 12$

8a. $p - 6 = 3$ b. $q - 4 = 5$

9a. $r - 1 = 100$ b. $s - 12 = 6$

10a. $a - 4 = 4$ b. $b - 100 = 1$

11a. $a + 10 = 16$ b. $f - 17 = 17$

12a. $b - 14 = 10$ **b.** $g - 41 = 11$

13a. $c + 12 = 24$ **b.** $h + 32 = 48$

14a. $d - 16 = 10$ **b.** $j + 16 = 61$

15a. $e + 18 = 41$ **b.** $x - 14 = 31$

Exercise 3

Write all of these questions as algebraic equations, then solve them:

1. Your friend thinks of a number. She adds 6 and tells you that the answer is 7. What number did she first think of?

2. Another friend thinks of a number, subtracts 10 and says the answer is 100. What was the first number?

3. Think of a number (n). Add 2. The answer is 10. Find n.

4. Think of a number (x). Subtract 2. The answer is 10. Find x.

5. Think of a number (a). Add 16. The answer is 32. Find a.

6. Think of a number (y). Add six. The answer is twelve. Find y.

7. Think of a number (p). Subtract twenty three. The answer is eleven. Find p.

8. Think of a number (c). Add seventeen. The answer is twenty nine. Find c.

9. Think of a number (k). Subtract eighteen. The answer is nine. Find k.

10. Think of a number (m). Add thirty seven. The answer if fifty. Find m.

Remember

Questions in algebra form are the easiest to work with.
1. Your friend thinks of a number. He multiplies it by 2 and tells you the answer is 10. What number did he first think of?
2. Think of a number (n). Multiply it by 2. The answer is 10. Find n.
3. Solve $2n = 10$.

The solutions to the questions are the same, 5, but the third is the easiest to work with.

Examples

1. Solve $4m = 20$
   $m = 5$

2. Solve $10t = 90$
   $t = 9$

Exercise 4

1a. $3a = 6$ **b.** $5z = 10$ **c.** $10g = 50$

2a. $2s = 4$ **b.** $2t = 20$ **c.** $6w = 24$

3a. $9y = 9$ **b.** $9d = 36$ **c.** $6f = 36$

4a. $5a = 35$ **b.** $6e = 54$ **c.** $7x = 21$

5a. $4u = 20$ **b.** $4c = 16$ **c.** $3q = 33$

6a. $5v = 50$ **b.** $3l = 15$ **c.** $3a = 99$

Write these questions as algebraic equations, then solve them:

7. 8 equal boxes of sweets contain a total of 80 sweets. How many sweets are there in each box?

8. Six equal circles have a combined area of forty eight square centimetres. Find the area of one circle in cm^2.

Examples

1. Solve $\dfrac{x}{4} = 2$

    $x = 8$

2. $\dfrac{a}{10} = 5$

    $a = 50$

Exercise 5

Solve each of the following equations:

1a. $\dfrac{a}{4} = 6$ b. $\dfrac{k}{2} = 20$ c. $\dfrac{g}{2} = 4$

2a. $\dfrac{x}{5} = 11$ b. $\dfrac{c}{6} = 6$ c. $\dfrac{q}{3} = 7$

3a. $\dfrac{r}{9} = 2$ b. $\dfrac{y}{5} = 9$ c. $\dfrac{z}{9} = 5$

4a. $\dfrac{d}{4} = 3$ b. $\dfrac{a}{6} = 8$ c. $\dfrac{w}{7} = 10$

5a. $\dfrac{a}{3} = 18$ b. $\dfrac{b}{6} = 13$ c. $\dfrac{c}{8} = 7$

Write each of these questions as algebraic equations, then solve them:

6. A friend thinks of a number. After dividing it by 2 he tells you the answer is now 20. What number did he first think of?

7. Another friend thinks of a number. After dividing it by 5 she tells you the answer is 11. What was the number she first thought of?

8. A box of table tennis balls was divided exactly between nine players. If each player received 6 balls, how many were there to begin with?

9. A packet of midget gems was shared between 12 girls. If each girl received 8 sweets, how many were there in the packet?

10. A box of biscuits was divided exactly between nine dogs. If each dog received 16 biscuits, how many were in the box to start with?

11. 192 tourists arrive at a beauty spot in four coaches. There are the same number of tourists in each coach. How many are there in each of the coaches?

12. A race of 10 000 m is broken up into laps, each of 500 m. How many laps are there?

Examples

1. Solve $3x + 1 = 10$

$$3x + 1 = 10$$
$$ - 1 -1$$
$$\text{so, } 3x = 9$$
$$x = 3$$

2. Solve $2c - 3 = 5$

$$2c - 3 = 5$$
$$ + 3 +3$$
$$2c = 8$$
$$c = 4$$

Exercise 6

Solve the following equations:

1a. $2s + 1 = 7$ **b.** $2w + 3 = 7$ **c.** $2a + 3 = 9$

2a. $3t + 4 = 7$ **b.** $3y + 3 = 12$ **c.** $4x + 6 = 30$

3a. $5a + 9 = 19$ **b.** $3a + 13 = 25$ **c.** $6b + 9 = 39$

4a. $7d + 18 = 39$ **b.** $7c + 13 = 41$ **c.** $3c + 4 = 16$

5a. $2q - 4 = 4$ **b.** $3p - 5 = 1$ **c.** $2d - 3 = 11$

6a. $4a - 10 = 2$ **b.** $10b - 10 = 10$ **c.** $8m - 6 = 74$

7a. $5k - 22 = 3$ **b.** $6u - 32 = 16$ **c.** $6p - 3 = 51$

8a. $2j - 2 = 0$ **b.** $5z - 5 = 5$ **c.** $2q - 16 = 4$

9a. $3a - 6 = 3$ **b.** $2c - 13 = 3$ **c.** $4s - 10 = 34$

10a. $2b + 10 = 14$ **b.** $4d + 21 = 29$ **c.** $3t + 14 = 50$

11a. $4p - 12 = 4$ **b.** $6e - 25 = 5$ **c.** $5x + 33 = 68$

12a. $2m - 14 = 2$ **b.** $8p - 6 = 18$ **c.** $7y - 21 = 28$

13a. $3p + 18 = 36$ **b.** $3m - 11 = 22$ **c.** $9k + 6 = 60$

14a. $5x - 15 = 10$ **b.** $9n - 14 = 13$ **c.** $6l - 18 = 24$

15a. $6y - 13 = 11$ **b.** $7p + 8 = 22$ **c.** $3m - 5 = 13$

16a. $7a - 22 = 6$ **b.** $5t - 8 = 17$ **c.** $2p + 15 = 51$

17a. $5b - 16 = 24$ **b.** $4k - 15 = 9$ **c.** $5s - 10 = 30$

18a. $6p + 14 = 20$ **b.** $3l + 10 = 28$ **c.** $7r - 12 = 9$

19a. $8r + 18 = 34$ **b.** $6m + 12 = 12$ **c.** $2t - 13 = 11$

20a. $3s - 19 = 11$ **b.** $8a - 16 = 16$ **c.** $8r + 9 = 73$

129 Algebra: expansion of brackets

Examples

1. $3(s + 2)$ is the same as $3s + 6$
2. $5(3c + 10)$ is the same as $15c + 50$
3. $2(4g - 3)$ is the same as $8g - 6$
4. $5(a - 1)$ is the same as $5a - 5$

Exercise 1

Expand each of the following:

1a. $3(2e + 3)$	b. $4(2q + 2)$	c. $6(3r + 1)$
2a. $5(4g + 4)$	b. $2(2w + 1)$	c. $8(3h + 9)$
3a. $4(d + 2)$	b. $4(w + 6)$	c. $3(a + 9)$
4a. $7(x + 10)$	b. $9(f + 1)$	c. $10(k + 10)$
5a. $3(2d - 5)$	b. $2(5t - 5)$	c. $10(3d - 9)$
6a. $6(y - 7)$	b. $3(a - 4)$	c. $4(10h - 4)$
7a. $17(e - 1)$	b. $12(s - 2)$	c. $9(v - 6)$
8a. $12(2a - 4)$	b. $10(3p - 5)$	c. $8(4b - 6)$

Examples

1. Solve $4(p + 2) = 16$

$$
\begin{array}{rcl}
4(p + 2) &=& 16 \\
\text{so, } 4p + 8 &=& 16 \\
-8 & & -8 \\
\hline
4p &=& 8 \\
p &=& 2
\end{array}
$$

Expand the brackets and then solve in the same way as before.

if $4(p + 2) = 16$, then $p = 2$

2. Solve $6(2w - 3) = 6$

$$
\begin{array}{rcl}
6(2w - 3) &=& 6 \\
12w - 18 &=& 6 \\
+18 & & +18 \\
\hline
12w &=& 24 \\
w &=& 2
\end{array}
$$

if $6(2w - 3) = 16$, then $w = 2$

Exercise 2

Solve the following equations:

1a. $4(a + 3) = 28$	b. $5(d + 4) = 40$	c. $3(y + 3) = 27$
2a. $5(r + 3) = 20$	b. $10(2w + 1) = 50$	c. $7(3f + 4) = 49$
3a. $9(c + 4) = 81$	b. $3(2a + 6) = 48$	c. $8(2x + 1) = 56$
4a. $2(w - 5) = 10$	b. $2(d - 2) = 4$	c. $6(t - 5) = 30$
5a. $2(2d - 2) = 4$	b. $3(4e - 2) = 6$	c. $6(3s - 3) = 0$

130 Algebra: solution of equations

Examples

1. If $6a + 24 = 4a + 28$, find a.
   Your aim is to end up with the a's on one side and the number on the other. You just subtract as shown.

   $$6a + 24 = 4a + 28$$

   subtract 4a from each side:
   $$-4a \qquad -4a$$

   $$2a + 24 = 28$$

   subtract 24 from each side:
   $$-24 \qquad -24$$

   $$2a = 4$$
   $$\text{so,} \quad a = 2$$

   if $6a + 24 = 4a + 28$, then $a = \mathbf{2}$

2. If $8x - 10 = 6x + 6$, find x.
   You do this example in exactly the same way as in the example above, but you must *add* the 10, because it was subtracted originally.

   $$8x - 10 = 6x + 6$$

   subtract 6x from each side:
   $$-6x \qquad -6x$$

   $$2x - 10 = 6$$

   add 10 to each side:
   $$+10 \qquad +10$$

   $$2x = 16$$
   $$\text{so,} \quad x = 8$$

   if $8x - 10 = 6x - 6$, then $x = \mathbf{8}$

Exercise 1

Solve the following equations:

1a. $6a + 12 = 3a + 18$ b. $7b + 10 = 5b + 20$

2a. $9n + 12 = 4n + 37$ b. $8m + 20 = 5m + 29$

3a. $3g + 12 = 2g + 28$ b. $7v + 10 = 3v + 30$

4a. $6a + 10 = 3a + 40$ b. $5s + 9 = s + 13$

5a. $7d + 14 = 5d + 22$ b. $6a + 12 = 5a + 13$

6a. $7y + 12 = 2y + 22$ b. $2w + 3 = w + 3$

7a. $12u + 1 = 3u + 10$ b. $10f = 5f + 5$

8a. $3e - 10 = 2e + 1$ b. $6h - 14 = 2h + 2$

9a. $5r - 12 = 2r + 12$ b. $4s - 4 = 2s + 10$

10a. $6t - 12 = 2t + 16$ b. $7p - 3 = 2p + 22$

11a. $8z - 9 = 3z + 26$ b. $16q - 16 = 10q + 2$

12a. $8a - 24 = 3a + 26$ b. $9w - 24 = 5w + 16$

13a. $14p - 13 = 5p + 5$ b. $18s - 20 = 8s + 50$

14a. $17x - 5 = 10x + 16$ b. $13b - 13 = 12b + 9$

15a. $9z - 10 = 3z + 50$ b. $11b - 21 = 6b + 4$

Algebra: **Revision test**

1. Simplify each of the following:
   a. $6a + 3a + 10a$
   b. $4x + 3y + 2y + 3x$
   c. $2p + 6q - 3q$
   d. $6a + 4b - 2a + 3b$
   e. $7a + 2b + 6c + 3a + 4b - 2c$
   f. $10p + 8q + 6r - 3q - 6p + 2r$

2. Use algebra to write the following in a shorter way:
   a. a plus 10
   b. sixteen minus double c
   c. p multiplied by q
   d. eleven subtracted from 3 times s
   e. x divided by four.

3. Rewrite the following in a shorter way:
   a. $a \times a$
   b. $b \times b \times b$
   c. $3 \times 3 \times p \times p$
   d. $4 \times w \times w \times 4 \times w$
   e. $d \times 6 \times d \times d \times 6$
   f. $2 \times 3 \times p \times 2 \times p$
   g. $y \times y \times 4 \times 8 \times y$
   h. $m \times 6 \times 2 \times m \times 2$
   i. $3 \times 3 \times 3 \times d \times d \times d$
   j. $3d \times 3d \times 3d$

4. Simplify the following:
   a. $3a \times 3 + 4a$
   b. $4 \times 6b - 10b$
   c. $2 \times 3p + 6 \times 2p$
   d. $4p \times 8 - 3 \times 6p$
   e. $2 \times 3p + 6 \times 4p$
   f. $2 \times 4a + 3 \times 5a - 6 \times 3a$
   g. $4b \times 7 - 2 \times 6b + 3 \times 5b$
   h. $2 \times 6 \times 3c + 4 \times 2c \times 3$
   i. $4y \times 3 \times 3 - 2 \times 3\, y \times 2$
   j. $3 \times 3 \times 2w$
   $+ 4w \times 3 \times 2 - 3 \times 2w \times 4$

5. Find the value of the following if $x = 5$ and $y = 3$:
   a. $2x + 4$
   b. $3y + 9$
   c. $x + y$
   d. $2y + 4x$
   e. $6x - 3y$
   f. x^2
   g. xy
   h. $2x^2 + x + 4$
   i. $3y^2 - 3y + 6$
   j. $(x + y)^2$

6. Write the following as algebraic equations, then use your equation to find the original number:
   a. Dave thinks of a number and adds eleven. The total is seventeen.
   b. Maureen thinks of a number and adds twelve. The total is twenty.
   c. Tony thinks of a number then adds thirteen. The total is twenty three.
   d. Joan thinks of a number then subtracts eight. The answer is three.
   e. Geoff thinks of a number then subtracts eighteen. The answer is thirteen.

7. Solve the following equations:
   a. $a + 2 = 16$
   b. $b + 6 = 11$
   c. $c + 11 = 16$
   d. $d - 14 = 2$
   e. $e - 19 = 6$
   f. $i - 21 = 4$
   g. $v + 25 = 35$
   h. $w - 30 = 6$
   i. $x + 40 = 52$
   j. $y - 50 = 16$

8. Solve:

   a. $2p = 14$
   b. $3q = 24$
   c. $6r = 18$
   d. $4s = 20$
   e. $5t = 35$
   f. $10l = 40$
   g. $9m = 63$
   h. $7n = 56$
   i. $8z = 40$
   j. $9k = 45$

9. Solve:

   a. $\dfrac{k}{3} = 7$
   b. $\dfrac{x}{2} = 4$

   c. $\dfrac{y}{6} = 6$
   d. $\dfrac{p}{4} = 10$

   e. $\dfrac{q}{6} = 12$
   f. $\dfrac{r}{5} = 6$

   g. $\dfrac{t}{8} = 8$
   h. $\dfrac{v}{7} = 9$

   i. $\dfrac{w}{4} = 10$
   j. $\dfrac{b}{2} = 11$

10. Solve:

    a. $3x + 6 = 15$
    b. $4y + 4 = 12$
    c. $2z - 3 = 9$
    d. $6l - 8 = 10$
    e. $7m + 7 = 14$
    f. $8n - 10 = 6$
    g. $4p - 12 = 20$
    h. $5q + 14 = 29$
    i. $6n - 11 = 13$
    j. $9s + 15 = 42$

11. Solve:

    a. $2(x + 3) = 12$
    b. $3(k + 6) = 30$
    c. $6(l - 4) = 12$
    d. $4(m - 2) = 12$
    e. $6(n + 6) = 48$
    f. $5(z + 3) = 60$
    g. $7(x - 1) = 63$
    h. $9(y - 4) = 27$
    i. $8(p + 6) = 56$
    j. $6(q + 7) = 54$

12. Solve:

    a. $2p + 10 = p + 14$
    b. $3q + 8 = 2q + 20$
    c. $4r - 10 = 2r + 16$
    d. $5s - 6 = 3s + 24$
    e. $6t + 13 = 2t + 17$
    f. $7v - 8 = 3v + 28$

Write these as algebraic equations, then solve them:

13. 4 tins of dog food + 3 tins of cat food cost 180p altogether. Dog food costs 30p per tin. How much does cat food cost?

14. The Murphy family (husband, wife, mother-in-law, and five children) go to the cinema. Admission for adults is £2·00. The total they pay to get in is £11. What is the admission charge for each child?

131 Frequency tables

Remember

Information is often correct, but hard to understand because it is not presented clearly. A *frequency table* can be used to help:

Example

Here are the scores from a test. The maximum possible score is 10:

```
5  9  3  6  2  3  3  2  5  7
6  1  5  4  6  2  4  2  3  3
0  3  6  3  3  4  0  8  1  2
3  3  2  1  4  1  7  4  4  1
```

These scores can be seen more easily using a *frequency table*: the table has been filled in for 0 and 4 marks out of 10:

marks out of 10	frequency tally	total
0	I I	2
1		
2		
3		
4	⊞ I	6
5		
6		
7		
8		
9		
10		

A tally-mark is made each time "0" occurs. Two people scored 0 out of 10.

The fifth tally-mark is done across the other four to make them easy to count.

Exercise 1

1. a. Copy and complete the frequency table, using the list of scores.
   b. Check that you have not missed any scores. Count the number of scores in the final list; then count the total number in your frequency table. The two totals should be the same.
   c. Do you think the test was easy or hard? Write a sentence to explain your answer.

2. The following marks were obtained by 60 pupils in a test. The maximum possible score was 10:

```
6  4  2  1  6  3  2  4  2  6  1  0
4  8  7  6  5  4  3  2  6  4  5  3
7  2  8  1  3  2  5  6  4  3  2  4
6  5  7  5  4  3  2  8  5  4  2  7
1  2  4  6  5  7  6  5  4  4  3  2
```

   a. Make a frequency table for the results of this test.
   b. Check that you have used all the data.
   c. Do you think the test was easy or hard? Write a sentence to explain your answer.

3. A stamp collector is given a packet of mixed stamps as a birthday
    present. The stamps came from Australia, Belgium, Canada, Denmark,
    Egypt, France and Germany. When he spread the stamps out on a
    table, he found that the countries of each of the stamps was as follows:

    A G F C E A A A G D
    G C F F F D B C A A   where 'A' stands for
    G F G F F D A A A A   Australia
    F F F F F B E G G F   'B' stands for
    A A G C G A F D C G   Belgium; etc.

    a. How many stamps was he given for his birthday?

    b. Copy and complete the following frequency table:

country	frequency tally	total
Australia		
Belgium		
Canada		
Denmark		
Egypt		
France		
Germany		

    c. Check that you have used all the data.

    d. Which country occurs most frequently in his present of stamps?

4. The list below gives the number of matches in each of a group of
    matchboxes:

49	51	50	49	54	52	48	49	51	53
48	46	52	50	51	47	49	52	53	54
50	49	54	51	46	49	48	50	54	53
52	49	50	52	51	50	48	46	49	50
54	52	49	47	48	52	51	48	50	52

    a. Copy and complete the following frequency table:

matches	frequency tally	total
46		
-		
-		
-		
-		
-		
-		
-		
54		

    b. Check the totals of the number of tally marks and the number of
    matchboxes given in the list to make sure that you have used all the
    data.

c. In this sample of matchboxes what number of matches occurred most frequently?

d. On each of these matchboxes was written 'Average Contents 50'. From this sample, do you accept that this is fair? Write a sentence to explain your answer.

5. Draw a frequency table to collect together the shapes below. Group them under: circles, triangles, quadrilaterals (4 sides) and pentagons (5 sides). Which shape is most frequent? Which is least frequent?

6. This list is of the percentages scored by pupils in an English examination:

24	38	59	41	27
86	91	56	62	47
28	26	63	53	26
48	53	29	60	80
22	96	46	36	28
23	72	41	40	10
61	53	47	62	32
43	62	82	50	49
33	17	22	33	7

score	tally	frequency
0- 9		
10-19		
20-29		
30-39		
40-49		
50-59		
60-69		
70-79		
80-89		
90-99		

a. Copy and complete the frequency table above. Grouping the scores like this makes it easier to interpret the information when there are many different numbers.

b. Do you think that the examination was easy or difficult? Give reasons for your answer.

132 Pie charts

Remember

A common way of showing information is to use a *pie chart*. A whole circle is drawn to represent the complete group. "Slices" are drawn which represent the relative size of each part of the group.

Example

A sample of twenty men on their way to work were looked at to see if they were wearing ties. Here is the information that was recorded:

	frequency	degrees
total number	20	360°
striped	6	108°
spotted	2	36°
plain	4	72°
no tie worn	8	144°

So the slice for each *person* will be
$360° \div 20 = 18°$
$(18° \times 6 = 108°)$

Using a protractor a piechart can be drawn:

Exercise 1

1. The table below gives the distances that a group of students must travel to get to school:

	less than 1 km	1-2 km	2-3 km	over 3 km
number of students	10	16	4	6

   a. What was the total number of students in the survey?
   b. What angle on a pie chart would represent one student?
   c. Complete a frequency table like the example above.
   d. Draw a pie chart to represent the information.

2. Out of a sample of 8 people, 4 are aged 10–19, 2 are aged 20–29, and the rest are aged 30 or over. Construct a pie chart to represent this data.

3. Here are the marks out of 10, for a group of 120 students:

2	8	6	7	6	5	4	6	7	8	9	6
5	3	5	6	4	3	5	9	8	6	5	5
6	4	3	2	6	2	6	7	8	9	8	7
6	5	4	3	4	6	8	9	5	6	5	9
7	6	2	4	8	6	3	4	8	6	5	2
7	4	2	6	3	6	4	8	5	4	2	3
7	8	9	5	4	5	6	8	6	5	6	5
6	3	4	5	2	6	5	4	2	5	6	8
4	2	5	6	7	8	5	6	5	6	5	4
3	4	5	5	4	2	3	7	9	7	8	9

   Make a frequency table and a pie chart from these results.

169

4. A schoolgirl made a survey on how she spent her time watching the television. She spent 65 minutes watching comedy programmes, 67 minutes watching music programmes, 9 minutes watching the news, 23 minutes watching quiz programmes and 16 minutes watching adverts.

    a. Make a table of this information.
    b. How many minutes has she spent watching television altogether? How many hours is this?
    c. Draw a pie chart to represent the information. What angle will represent each minute of watching time on the pie chart?
    d. Are you able to describe the schoolgirl's likes and dislikes from her survey?

133 Pictographs

Remember   Advertisements often show numerical information using a *pictograph*:

Example   The pictograph below shows the amount of traffic passing a bus stop in a five minute period:

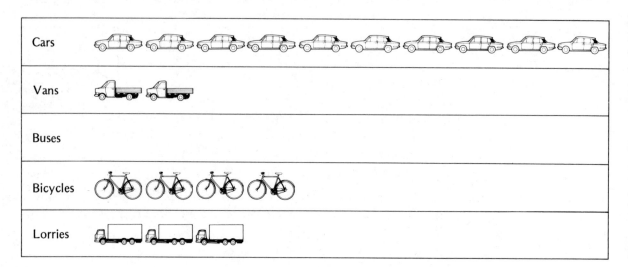

Cars	
Vans	
Buses	
Bicycles	
Lorries	

Exercise 1

1. a. How many cars passed the bus stop in this 5 minute period?
    b. How many bicycles passed the bus stop?
    c. How many buses passed the bus stop?
    d. Can you think of the reason why the vehicles have all been drawn the same length rather than to the same scale?

2. The pictograph below represents the number of cups of tea drunk by one person during the week at work.

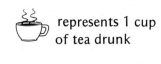 represents 1 cup of tea drunk

a. How many cups of tea did this person drink on Monday?
b. How many cups of tea did this person drink on Friday?
c. On which day were most cups of tea drunk?
d. What was the total number of cups drunk during the week?
e. If tea was charged at 7p a cup, how much did this person spend on tea that week?

3. The pictograph below shows the number of cups of tea drunk by the staff of a school in the course of a week. The staff of the school are divided up into departments. *Each picture of a cup represents 10 cups of tea.*

English	🍵 🍵 🍵 🍵 🍵 🍵 🍵 🍵 🍵 🍵
Mathematics	🍵 🍵 🍵 🍵 🍵 🍵 🍵 🍵 🍵 🍵 ½
Science	🍵 🍵 🍵 🍵 🍵 🍵
Social Studies	🍵 🍵 🍵 🍵 🍵 🍵 🍵 🍵
Languages	🍵 🍵 🍵 🍵 🍵 🍵 🍵
Religious Studies	🍵 ½

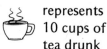 represents 10 cups of tea drunk

a. How many cups of tea did the English department drink during the week?
b. What does the 'half cup' represent in the mathematics department rating?
c. How many cups of tea did the mathematics department drink during this week?
d. Is it possible to say from this pictograph that a teacher in the mathematics department drank more tea in this week than a teacher in the religious studies department?

134 Bar charts

Information is often shown accurately and simply using a *bar chart*:

Here is some information about rainfall during one week:

day	Mon	Tues	Wed	Thur	Fri	Sat	Sun
rainfall (cm)	1	2	1	1.5	0	0	2.5

This can be shown on a bar chart. Monday and Tuesday have been drawn in:

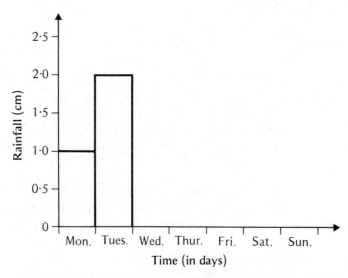

1. a. Copy the bar chart above and complete it for the rest of the week. Remember to keep the *width* of each bar exactly the same.
   b. Which days do you think would have been good for sunbathing?
   c. What was the total amount of rain during the week?

2. The table below gives approximate figures for Britain's exports, as a percentage of the total value of exports.

type of export	road vehicle	metals	food, drink	machinery	chemicals	other
percentage	9%	10%	7%	28%	10%	36%

Draw a bar chart to display this information.

3. The bar chart below gives a comparison between the maximum and minimum temperatures on 6 consecutive days.

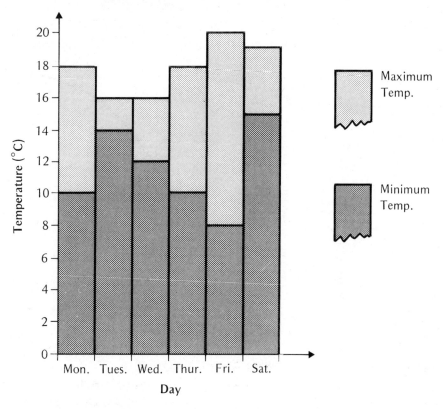

a. Which day had the lowest temperature of the week?
b. Which day of the week had the highest temperature?
c. Which day of the week had the highest minimum temperature?
d. Which day of the week do you think would have been most pleasant to sit outside? Give reasons for your answer.
e. What time of the year do you think that these statistics were taken from?

4. Look back at the frequency tables you made in section 131. Draw suitable bar charts for each question:

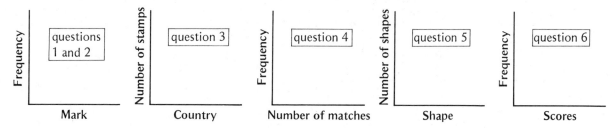

135 Line graphs

Remember A *line graph* is the best type of graph to use, if you are measuring something which is *gradually changing*.

Example The graph below represents the height of a plant from seed, over a period of 16 days:

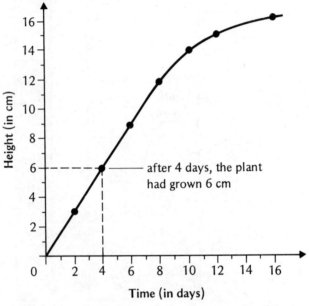

after 4 days, the plant had grown 6 cm

Exercise 1

1. a. What was the height of the plant on the 6th day?
    b. What was its height on the 8th day?
    c. What was its height (approximately) on the 9th day?
    d. What was its height on the 16th day?
    e. How often was the plant actually measured?

2. Look carefully at the graph below: it is a graph of the temperature of water after it was placed in the freezing compartment of a fridge to make ice. The temperature was taken every 10 minutes.

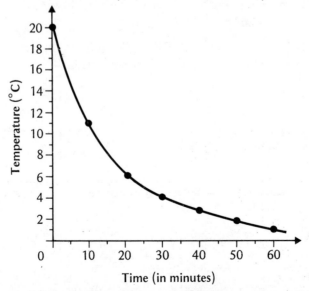

a. What was the temperature of the water 10 minutes after it was placed in the freezing compartment?

b. What was the temperature of the water 30 minutes after it was placed in the freezing compartment?

c. How long was it before the water was at 1°C?

d. Approximately how long was it before the water was at 5°C?

e. Can you tell from the graph what the approximate freezing point of water is? If you can, write down what you think it is.

f. Fill in the rest of this table, using information from the graph:

time after being put in freezing compartment (min.)	0	10	20	30	40	50	60
approximate temperature (°C)							

g. Guess what the temperature would be after 70 minutes.

3. The table below gives the temperatures of water in a kettle allowed to cool after it has boiled.

time after boiling (min.)	0	2	4	6	8	10
temperature (°C)	100	50	30	25	22	21

a. Plot the points given by the information above. Use a graph like this:

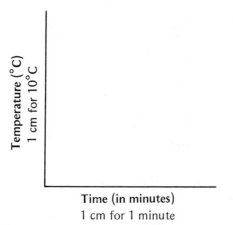

Temperature (°C)
1 cm for 10°C

Time (in minutes)
1 cm for 1 minute

b. Is it reasonable to join the points with a line? Give reasons for your answer.

c. What do you think was the temperature of the room where the water was allowed to cool?

d. Can you say from your graph what the temperature of the water was after 1 minute, approximately? If so, write down your answer.

4. The table below shows the *percentage* of people in Britain who lived in towns, over the 150 years up to 1950:

year	1800	1825	1850	1875	1900	1925	1950
percentage living in towns	20	30	40	50	60	70	75

Note:
In this table a town is regarded as being a place with more than 10,000 inhabitants.

a. Plot the points given by the information above. You should have the year along the bottom and the percentage living in towns up the side.
b. Is it reasonable to join the points with a line? Give reasons for your answer.
c. Can you predict from the graph roughly what percentage of the population will live in a town in the year 2000? If you think that you can predict this, write down your prediction.
d. Can you tell from the graph roughly what percentage of the population lived in towns in 1775? If so, write down what you think it was.

5. The graph below represents the journey of a train, for a period of 20 minutes, after leaving a station.

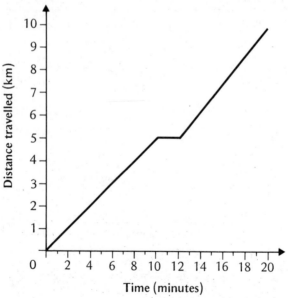

a. How far did the train travel in the first six minutes?
b. How far did the train travel in this period of 20 minutes?
c. How long did the train take *to travel the* first *5 km?*
d. How long did the train take *to travel the* last *5 km?*
e. Between which times was the train travelling fastest?
f. What was happening between the 10th minute and 12th minute, do you think?
g. What did the driver do to make up the lost time?

6. The table below gives the distances that a car has travelled at various times after it has left a set of traffic lights.

time (minutes)	0	2	4	6	8	10	12	14	16	18	20
distance travelled (km)	0	1	2	3	3	3	3	4	8	8	8

a. Plot the points given by the information above on a distance/time graph.

b. How far did the car travel in the first 4 minutes?

c. How far did the car travel in the first 5 minutes? Are you able to say for *certain* that this was the distance travelled in the first 5 minutes?

d. Write some story that could describe the journey of the car as it is shown on the graph.

136 Cheating with graphs

Remember

Graphs can be used to show information clearly. They can also be used to give the *wrong* impression, by "chopping off" part of the scale:

Example

The two graphs below represent the increase in a student's weekly allowance over a period of six months:

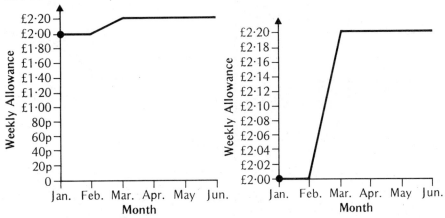

"Fair enough. It's time you had an increase!"

"You've had such an increase this year, you're not getting any more!"

Exercise 1

1. a. Compare the two graphs. What was the student's allowance at the beginning of June? What was it at the beginning of the year?

b. How much had the allowance gone up over the half-year altogether?

c. Which graph would *you* use as 'support' if you wanted an increase in allowance?

d. Both graphs show exactly the same information. What makes them look different?

e. Which graph do you think is fairer?

177

2. These graphs show the results of a test on a car fitted with Kwikgrab brakes:

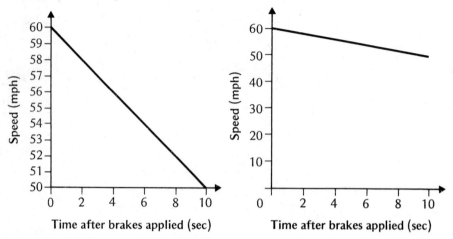

a. The two graphs above represent the same information. Explain why they look so different.

b. Which graph gives the truer picture of the quality of Kwikgrab brakes?

c. If you were an advertising executive at Kwikgrab, which graph would you use? Why?

3. The graph below represents a rise in temperature over the period of a day:

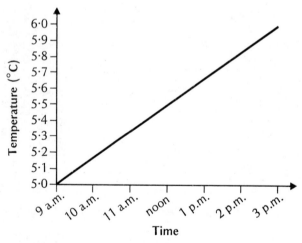

"Boy has it got hot today!"

a. Criticise the graph above, and the statement accompanying it.

b. What has been the increase in temperature over the day?

c. Draw a graph to show in a fairer way, the increase in temperature over the day.

4.

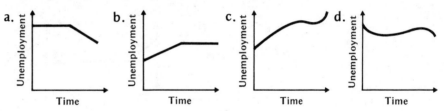

INVEST WITH US AND~

Double your money every 10 years

£100

£200 after 10 years

£400 after 20 years

← 1 cm →   ← 2 cm → ←————4 cm————→

The pictures above have been drawn exactly to scale. But do they give a fair picture of the growth of the money? Give reasons for your answer.

In questions 5 to 12 below, you are given a statement and a selection of graphs. You must choose the graph that *best* fits the statement.

5. "Unemployment is going up — again!"

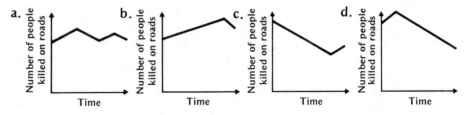

6. "After a steady increase over the last year, the number of people killed on the roads this month has dropped."

7. "The cost of bread has remained steady over the past year."

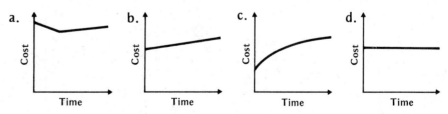

8. "The cost of bread has increased steadily this year."

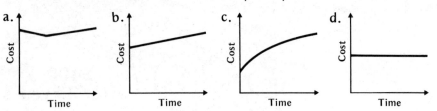

9. "The increase in the cost of bread in the shops has been less marked so far this year than it was last year."

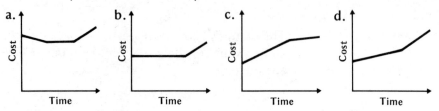

10. "After a poor start, our car sales have increased tremendously this summer."

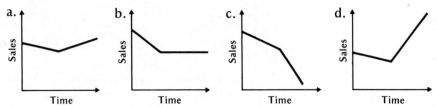

11. "Football attendance started well, then slumped during the winter but has almost picked up again."

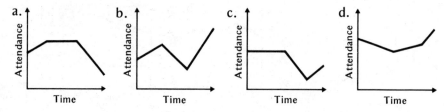

12. "Athletes expenses have increased considerably over the season."

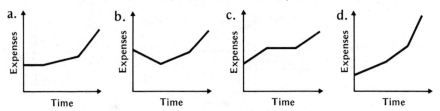

137 Averages: the mean

Remember

The "average" of a set of numbers is more correctly known as the *arithmetic mean*, or simply, *the mean*.

Examples

1. Find the mean of 6p, 15p and 3p.

$$\text{the mean} = \frac{6p + 15p + 3p}{3}$$
$$= \frac{24p}{3}$$
$$= \mathbf{8p}$$

2. A cricket batsman scored 16, 24, 2 and 6 in four matches. Find the mean of his scores.

$$\text{the mean of his scores} = \frac{16 + 24 + 2 + 6}{4}$$
$$= \frac{48}{4}$$
$$= \mathbf{12}$$

Exercise 1

Find the mean of the following:

1a. 6, 8, 12 and 2   b. 4, 8, 3 and 9

2a. 14, 18, 25 and 7   b. 26, 31, 17, 10 and 11

3a. 56, 86, 30 and 12   b. 1, 1, 1, 1, and 1

4. The scores of a batsman in 6 matches, which were 41, 86, 106, 8, 3 and 8.

5. The number of points scored by a netball player in 5 games, which were 38, 16, 10, 14 and 2.

6. The number of points scored by a table-tennis player in 6 games: 21, 18, 21, 22, 10 and 10.

7. The scores of a darts player in 9 throws of one dart: 20, 1, 40, 20, 5, 0, 40, 15 and 3.

8. The number of points scored by a basketball player in 5 games: 14, 16, 4, 22 and 24.

9. The attendances at 5 Bradford City's home League games: 2585, 2324, 1916, 3017 and 2218.

10. The times that it takes you to get to school on the mornings of one week. Time yourself each morning, record the times and then find the mean time for your journey.

138 Mode and median

Remember

As well as the mean, there are two other types of average, the *mode* and the *median*. However, the mean is quite the most important and is the type of average that people usually refer to when giving 'an average'. The *mode* is sometimes used when there are many items that give the same score — it is the most frequently occuring number.

Examples

1. Find the mode of 1, 4, 5, 3, 2, 3, 3, 6, 4, 3 and 3.
First, list the numbers in order. Then choose the most frequently occuring number:
1, 2, 3, 3, 3, 3, 3, 4, 4, 5, 6.
The mode is 3.

The *median* is sometimes used when most numbers are close to each other, with a few that are much much larger or smaller than this group — it is the *middle* number.

2. Find the median of 1, 4, 5, 3, 4, 3, 3, 6, 4, 3 and 16.
First, list the numbers in order. Then find the middle number:
1, 3, 3, 3, 3, 4, 4, 4, 5, 6, 16
The circled dot represents a total score of $(4 + 3) = 7$. Look at the list), so **the median is 4.**

Exercise 1

First, list the numbers in order. Then find the mode:

1a. 5, 4, 7, 1, 3, 1, 1, 5, 4   b. 4, 6, 4, 2, 7, 4, 1, 4, 3

2a. 2, 3, 4, 3, 5, 3, 7, 1, 3   b. 2, 2, 2, 5, 1, 0, 1, 5, 1, 5, 8, 5, 3

3a. 19, 30, 14, 19, 18, 19, 29   b. 130, 131, 132, 129, 130, 131, 130

Exercise 2

1. Find the median of each of the sets of numbers in exercise 1.

2. 2, 2, 6, 3, 2, 3, 9, 5, 4, 3, 2, 8, 9, 6, 2, 4, 5, 4, 4, 6, 1, 5, 2, 5, 3, 0, 9, 7, 8, 3, 2, 4, 9, 6, 2, 7, 5, 8, 3, 2, 6, 2, 2, 9, 3, 7, 4, 0.
   a. Compile a frequency table for the information above.
   b. From your frequency table, give the mode of these numbers.
   c. Find the median of the numbers.

3. A survey on the heights (in metres) of a local police force was as follows:

1·80	1·81	1·84	1·85	1·84	1·81
1·83	1·81	1·80	1·84	1·85	1·82
1·81	1·83	1·84	1·81	1·80	1·82
1·83	1·84	1·83	1·80	1·81	1·84
1·81	1·85	1·80	1·81	1·84	

   a. construct a frequency table.
   b. find the median.
   c. find the mode.

139 Mean, median and mode

Remember The median and mode do not only apply to numbers.

Example Find the mode and median of these letters: B C A D A C C B C B A
Re-arrange: A A A B B B C C C C D   C is the mode, B is the median

Exercise 1

1. A C D E B D E B C A D E
    B E C B A C E B C A B C
    C B A B C A B C B A E B
    C C B B C D E B B C B C
    E A B C B C D E A B C

    a. Compile a frequency table for the information above.
    b. Write down the modal letter.
    c. Write down the median of the letters.
    d. Is it possible to find an arithmetic mean for the letters above? Give the reasons for your answer.

2. The following scores give percentages scored in a test.

    6   10   29   76   83   96   57   60   24   24   64   57   83   65
    61   60   94   17   36   93   63   55   73   47   44   86   26   94
    29   47   75   21   56   6   47   16   83   40   53   43   36   38
    20   56   64   57   67   30   21   53   27   50   63   71   63

    a. Copy and complete the following frequency table for the scores above:

score	frequency
0– 9	
10–19	
20–29	
30–39	
40–49	
50–59	
60–69	
70–79	
80–89	
90–99	

    b. Write down the modal group of scores.
    c. Find the median group of scores.
    d. Is it possible to find the mean of these scores? Give the reasons for your answer.

3. Over a full season a top class athlete recorded the following positions in his races:

    2, 3, 6, 1, 1, 4, 5, 1, 2, 6,
    3, 2, 4, 3, 1, 2, 3, 4, 6, 1,
    2, 3, 4, 1, 1, 3, 2, 1, 3, 2

    Make a frequency table and find the mode.

140 Probability

Remember
A *possibility space* is a way of showing on paper all the combinations of events that can happen.

Example
Two dice are thrown at the same time and their scores are added. One die is green and the other is red.
The table below represents all the possible scores, and all the different ways that you can get them. It is called a *possibility space*, because it has every possible throw represented on it:

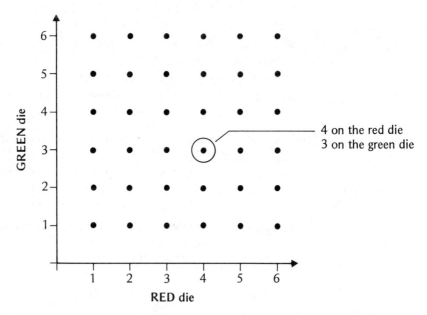

4 on the red die
3 on the green die

Exercise 1

1. What is the highest possible total score?

2. What is the lowest possible total score?

3. The circled dot represents a total score of (4 + 3) = 7. Look at the possibility space and write down all the scores that give a total score of 7. You should have six scores written down altogether.

4. Copy the possibility space and draw a 'loop' to surround all the points that give a total of 7.

5. On the same possibility space draw a loop to surround *all* the points that represent a 'double', for example, 3 on the red die and 3 on the green die.

6. Is it possible to throw *both* a double *and* a total score of 7 with one throw of the dice? Check on your possibility space that you don't have any points that are in *both* your loops.

7. Look at the two loops on your possibility space and find out how many ways there are to throw *either* a double *or* a total of 7.

8. How many possible results are there *altogether* when you throw the two dice?

141 Probability: single event

Remember

When you fill in a football pools coupon, or bet on a horse or dog, you are gambling in the hope of winning money. The chances of winning are not good and most gamblers lose. But they keep on trying in the hope of a big win. The chance of succeeding (or failing) is the *probability* of your success (or failure).

Examples

1. Find the probability of throwing a six with one die:

$$P \text{ (throwing a 6)} = \frac{1}{6} = \frac{\text{(the number of 6's on die)}}{\text{(the total number of faces)}}$$

2. Find the probability of picking an ace out of a pack of 52 cards:

$$P \text{ (picking out an ace)} = \frac{4}{52} = \frac{\text{(the number of aces)}}{\text{(the number of cards)}}$$

$$= \frac{1}{13}$$

From the examples you should be able to see the formula for finding the probability of a single event:

$$\text{Probability of an event} = \frac{\text{number of times that event can occur}}{\text{total number of possible events}}$$

Exercise 1

Find the probability of each of the following events:

1. Throwing a 4 with a throw of one die.

2. Picking out a King from a pack of cards.

3. Picking out a spade from a pack of cards.

4. Picking out the ace of spades from a pack of cards.

5. Tossing a coin and getting a head.

6. Throwing an even number with one die.

7. Picking out a King from a pack of cards that is lacking the 7 of spades.

8. Picking out a 7 from a pack of cards that is lacking the 7 of spades.

9. Picking out a red card from a pack of cards that is lacking the 7 of spades.

10. A bag contains 6 nasty sweets and 8 nice ones. What is the probability of picking out a nice one, assuming that you aren't allowed to look in the bag?

11. A box contains 12 blue socks and 8 red ones. What is the probability of choosing a red one?

12. A pond contains 16 trout and 4 salmon. Find the probability of catching a trout, assuming that it is just as easy to catch both types of fish.

142 Probability: combinations of event

Remember

If there is more than one way of winning, you are more likely to do so, provided that the number of ways of losing has not increased as well! So the probabilities of the separate events are *added*.

Examples

1. A game is played so that you win if either a 5 or a 6 comes up when the die is thrown. What is the probability of you winning?

P (winning) $= P$ (throwing a 5 *or* a 6)

$\qquad\qquad = P$ (throwing a 5) $+ P$ (throwing a 6)

$$= \quad \frac{1}{6} \quad + \quad \frac{1}{6}$$

$$= \frac{2}{6} \quad = \frac{1}{3}$$

The separate probabilities have been *added*. As a general rule, when you see the word '*or*' (here, as '5 *or* a 6'), *add* the separate probabilities.

2. Find the probability of picking a Queen or a King from a full pack of cards.

P (picking a Q or a K) $= P$ (a Queen) $+ P$ (a King)

$$= \quad \frac{4}{52} \quad + \quad \frac{4}{52}$$

$$= \frac{8}{52} \quad = \frac{2}{13}$$

Exercise 1

Find the probability of each of the following combinations of events:

1. Throwing a 2 or a 3 with one die.

2. Throwing a 2 or a 3 or a 4 with one die.

3. Throwing a 6 or a 1 with one die.

4. Picking a spade or a heart from a full pack.

5. Picking a spade or a heart from a pack that is missing the Queen of Diamonds.

6. A bag contains 12 black scarves, 4 red scarves and a green one. Find the probability of picking out (in one try)
   a. a black or a green scarf
   b. a red or a black scarf
   c. a green or a red or a black scarf

7. There are 12 knives, 6 spoons and 12 forks in a drawer. A boy picks one of them out without looking. Find the probability that:
   a. he chose a fork or a knife
   b. he chose a spoon or a fork.
   c. he chose a knife or a fork.
   d. he chose a fork or a spoon or a knife.

Remember

If two or more unlikely things happen before you win, the probability of winning gets less. Since the separate probabilities of each are always less than 1, *multiplying* the probabilities has this "further lessening" effect:

Examples

1. A game is played so that you win if, when a die is thrown twice, you get a 5 the first time *and* a 4 the second time.

    P (winning) $= P$ (throwing a 5 *and* a 4)

    $\qquad\qquad = P$ (throwing a 5) $\times P$ (throwing a 4)

    $\qquad\qquad = \quad \dfrac{1}{6} \quad \times \quad \dfrac{1}{6}$

    $\qquad\qquad = \dfrac{1}{36}$

    The separate probabilities have been *multiplied*. As a general rule, when you see the word '*and*' (here, as '5 *and* a 4') *multiply* the separate probabilities.

2. A card is taken from a pack, replaced, and then another card is taken from the same pack of cards. Find the probability that a King was taken out first, followed by a Queen.

    P (King and a Queen) $= P$ (King) $\times P$ (Queen)

    $\qquad\qquad = \dfrac{4}{52} \times \dfrac{4}{52} = \dfrac{1}{13} \times \dfrac{1}{13}$

    $\qquad\qquad = \dfrac{1}{169}$

Exercise 2

Find the probabilities of each of the following combinations of events.

1. Throwing a 6 and then a 4 in two throws of a die.

2. Throwing a 1 and then another 1 in two throws of a die.

3. Getting a head and then another head in two tosses of a coin.

4. Getting a head and then a tail in two tosses of a coin.

5. A bag contains 6 green balls and 3 red balls. A ball is taken out and then replaced. Another ball is then taken out. Find the probability of taking out a green ball the first time and the second time.

6. A small village pond contains 10 guppies, 6 sticklebacks and 4 minnows. An angler catches a fish, throws it back and then catches another. If the probability of catching each type of fish is the same, find the probabilities that he catches:
    a. a minnow and then a guppy
    b. a stickleback and then a guppy
    c. a guppy and then another guppy

7. Three books are on a shelf. One is green, one is red and the other is blue. You reach up and take one down without looking. You *don't* replace it, and then take down a second book, again without looking. Find the probability that the first book was blue and the second one red.

Graphs and statistics: **Revision test**

1. A farmer collected his eggs one morning and graded them from
   1–6 according to size. Here is his collection:

1	4	5	3	4	4	3	5	3
2	4	2	4	5	6	4	3	4
4	6	3	6	3	2	3	1	4
5	5	6	1	3	4	1	5	3
6	2	1	4	2	3	4	5	2

   a. Draw a frequency table.
   b. Which size of egg was most frequent?
   c. Draw a pie chart to represent the information.

2. A party of children go on a day's outing. Here is a bar chart showing
   how much pocket money they take with them:

   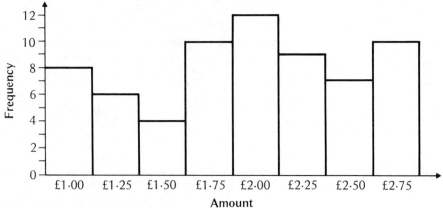

   a. How many pupils were on the outing?
   b. How many pupils took £2·50?
   c. How many pupils took more than £2 pocket money?
   d. How much money was taken altogether by the party?

3. The line graph below shows the number of words per minute that a
   typist is able to type, over a period of 4 weeks.

   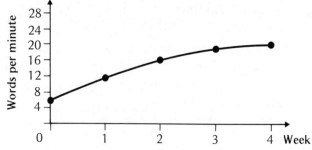

   a. How many words per minute could the typist type at the end of
   the first week?
   b. How many words per minute could the typist type at the end of the
   fourth week?
   c. How long do you think that the typist had been typing for when this
   graph was started?

4. Match each statement with one of the graphs shown:

A. Bread prices are steadily increasing.

B. After a bad start, car sales are much better.

C. We started the year well but things have got worse.

D. It has been an up and down year.

E. Nothing has changed at all.

1. 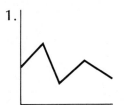 2. 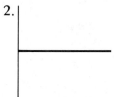 3. 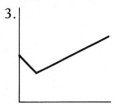 4.  5.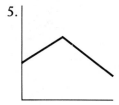

5. Find the arithmetic mean of the following numbers:

4   8   6   5   7   6.

6. A batsman scored the following number of runs in five successive innings. Find his average score:

64   41   86   32   82.

7. A survey was carried out on how many cups of tea were drunk by teachers in a school each day. The results are shown below:

```
4   8   5   2   1   6   7   4   2
6   6   2   6   1   5   0   2   1
2   4   1   5   4   0   3   3   4
1   9   4   3   6   4   0   5   3
3   5   7   2   3   1   4   7   2
```

a. Make a frequency table.
b. Draw a bar chart.
c. Find the modal number of cups of tea.

8. What is the probability of:

a. throwing an odd number with one die?
b. picking a red card from a complete pack?
c. tossing a coin and getting a tail?
d. A bag contains 8 red and 5 blue marbles. What is the probability of picking out a red?

9. A drawer contains 6 black socks, 8 white, 4 brown and 12 blue. Find the probability of choosing:

a. a brown sock
b. a brown or blue sock
c. a black or white sock
d. a black or brown or blue sock
e. a red sock.

10. At a school fete, you win a car if you can throw six dice (in one throw) and they all score six. What is the probability of this happening?

189

Remember   Most of the exercises in this book can be done without a calculator.
But it is often helpful if you can use a calculator quickly and accurately.

Remember   After switching on your calculator,
always press the AC button to
clear off any previous work.

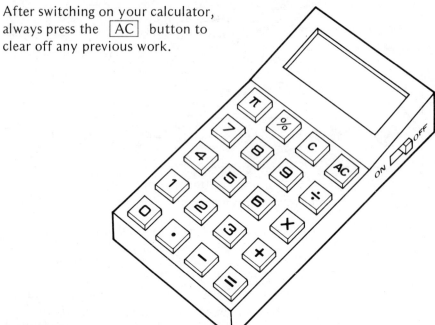

Addition

Examples

1. Calculate 62 + 123 + 86:

    |AC| |6||2| |+| |1||2||3| |+| |8||6| |=| **271**

2. Calculate 8·6 + 0·04 + 121·4

    |AC| |8||·||6| |+| |0||·||0||4| |+| |1||2||1||·||4| |=| **130·04**

Exercise 1

1a. 62 + 18   b. 28 + 36   c. 42 + 65 + 32

2a. 124 + 231   b. 231 + 106 + 308   c. 118 + 427 + 531

3a. 8·6 + 2·4   b. 12·9 + 16·8   c. 2·44 + 0·78

4a. 16·44 + 26·22   b. 11·8 + 21·2 + 16·5   c. 14·7 + 18·6 + 3·4

5a. 2·66 + 3·82 + 1·46   b. 16·6 + 8·26 + 1·24   c. 0·03 + 0·15 + 0·97

Subtraction

Examples

1. Calculate 124 − 86:

    |AC| |1||2||4| |−| |8||6| |=| **38**

2. Calculate 16·4 − 8·66:

    |AC| |1||6||·||4| |−| |8||·||6||6| |=| **7·74**

Exercise 2	**1a.** 19 − 8	**b.** 20 − 16	**c.** 26 − 9
	2a. 231 − 175	**b.** 425 − 197	**c.** 19073 − 4639
	3a. 8·6 − 4·4	**b.** 10·8 − 6·9	**c.** 14·6 − 8·8
	4a. 9·67 − 3·22	**b.** 11·62 − 9·88	**c.** 24·66 − 14·38
	5a. 68·21 − 35·35	**b.** 126·4 − 59·38	**c.** 234·9 − 78·34

Multiplication

Examples

1. Calculate 64 × 41:

 2624

2. Calculate 8·6 × 4·9:

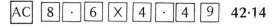 42·14

Exercise 3	**1a.** 9 × 8	**b.** 13 × 13	**c.** 52 × 121
	2a. 66 × 133	**b.** 199 × 99	**c.** 1478 × 297
	3a. 3·4 × 4·2	**b.** 4·9 × 6·8	**c.** 12·4 × 13·6
	4a. 14·8 × 28·6	**b.** 22·4 × 28·6	**c.** 9·66 × 0·42
	5a. 0·66 × 3·86	**b.** 8·27 × 3·75	**c.** 0·28 × 0·69

Division

Examples

1. Calculate 126 ÷ 86:

|AC| |1||2||6||÷||8||6||=| 1·4651163 = **1·465 to 3 d.p.**

2. Calculate 16·8 ÷ 3·2:

|AC| |1||6||·||8||÷||3||·||2||=| **5·25**

When sums do not work out exactly, record them to 3 decimal places.

Exercise 4	**1a.** 44 ÷ 11	**b.** 132 ÷ 12	**c.** 360 ÷ 15
	2a. 644 ÷ 126	**b.** 843 ÷ 131	**c.** 764 ÷ 236
	3a. 8·6 ÷ 4·3	**b.** 12·4 ÷ 3·1	**c.** 51 ÷ 7·5
	4a. 12·1 ÷ 181·5	**b.** 14·6 ÷ 292	**c.** 18·5 ÷ 496
	5a. 674 ÷ 0·6	**b.** 0·03 ÷ 0·0041	**c.** 0·96 ÷ 0·996

Copy this crossword into your book. Note the pattern in the squares!
Then fill in the answers:

ACROSS

1. $22 \cdot 6 - 10 \cdot 6$

2. $136 \div 2$

4. $(52 \div 4) \times 3$

6. 64×23

8. $375 \div 15$

10. $89 + 151 + 86$

12. $2000 \div 16$

14. 40×16

15. $500 - 149 - 101$

16. $391 \div 17$

19. 20×200

21. $16 + 24 - 13 - 9$

22. $900 \div 15$

23. $5 \cdot 5 \times 12$

DOWN

1. $64 - 51$

2. $122 + 264 + 256$

3. $1000 - 125$

5. $9 \cdot 3 \times 10$

7. $650 + 415 + 175$

9. 130×25

10. $864 \div 24$

11. $200 - 65 - 75$

12. $3 \cdot 4 \times 2 + 5 \cdot 2$

13. $63 + 26 - 39$

16. $15 \times 12 + 26$

17. $1 \cdot 5 \times 200$

18. $364 \cdot 5 \div 4 \cdot 5$

20. $653 \cdot 6 \div 7 \cdot 6$